Quantitative and Empirical Analysis of
Energy Markets

World Scientific Series on Energy and Resource Economics
(ISSN: 1793-4184)

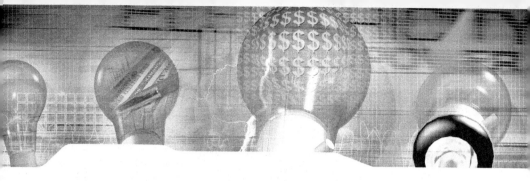

World Scientific Series on Energy and Resource Economics – Vol. 1

Quantitative and Empirical Analysis of
Energy Markets

Apostolos Serletis
University of Calgary, Canada

 **World Scientific**

NEW JERSEY · LONDON · SINGAPORE · BEIJING · SHANGHAI · HONG KONG · TAIPEI · CHENNAI

Published by

World Scientific Publishing Co. Pte. Ltd.

5 Toh Tuck Link, Singapore 596224

USA office: 27 Warren Street, Suite 401-402, Hackensack, NJ 07601

UK office: 57 Shelton Street, Covent Garden, London WC2H 9HE

Library of Congress Cataloging-in-Publication Data
Serletis, Apostolos.
    Quantitative and empirical analysis of energy markets / by Apostolos Serletis.
      p. cm. -- (World Scientific series on energy and resource economics ; vol. 1)
    Includes bibliographical references and index.
    ISBN 978-981-270-474-0 (alk. paper)
    1. Energy industries--Econometric models. I. Title.

    HD9502.A2S4525 2007
    333.7901'.5195--dc22   2007005203

British Library Cataloguing-in-Publication Data
A catalogue record for this book is available from the British Library.

First published 2007 (Hardcover)
Reprinted 2016 (in paperback edition)
ISBN 978-981-3203-38-9

Printed in Singapore

Contents

Foreword

This important book presents nineteen chapters of econometric time series analysis of crude oil, natural gas, and electricity markets. The economic structure of the energy markets is rapidly evolving, with the electric markets in many countries being deregulated. In Canada the prices and quantity supplied of electricity in the provinces of Ontario and Alberta are determined by a spot market; the electricity market in the United Kingdom has been deregulated for a number of years. Thus it is possible to create empirical studies of the evolution of these electricity markets.

The electricity markets chapters in this book concentrate on the North American market. The lessons learned from the empirical studies presented in this book can serve as a guide for planning electricity deregulation in the United States, the European Union, and Australia.

The electricity markets are related to the oil and gas markets, since electricity can be generated by burning gas or oil or coal depending on the technology of each power plant in the grid. There are several chapters in the book that present empirical results about the interrelations of the electricity, natural gas, and oil markets in North America.

The unique and important methodological contribution in several of these chapters is the use of nonlinear time series methods to study the nonlinear nature of the energy spot and futures markets. Although it is now well known that the economic system is nonlinear, the standard approach to studying markets is to employ linear time series methods. Linear models are adequate for aliased monthly and quarterly time series, but they are too crude for high frequency data.

Serletis and his coauthors employ more sophisticated nonlinear methods to the study of market volatility than the popular ARCH and GARCH models, which are known to have poor forecasting properties.

I consider this book to be a template for future econometric studies of the evolution of the dynamics of the energy market. What is now needed is a synthesis of the engineering, economics, political and legal aspects of a deregulated global energy market.

Melvin J. Hinich

Mike Hogg Professor, Department of Government,
Professor of Economics,
and
Research Professor, Applied Research Laboratories,
The University of Texas at Austin,
Austin TX 78713-8029

Part 1

Crude Oil Markets

Overview of Part 1

Apostolos Serletis

The following table contains a brief summary of the contents of each chapter in Part 1 of the book. This part of the book consists of five chapters dealing with recent state-of-the-art advances in the field of applied econometrics and their application to petroleum prices.

Crude Oil Markets

Chapter Number	Chapter Title	Contents
1	Unit Root Behavior in Energy Futures Prices	This chapter tests for random walk behavior in crude oil, heating oil, and unleaded gas futures prices and shows that the random walk hypothesis can be rejected if allowance is made for the possibility of a one-time break in the intercept and the slope of the trend function.
2	Rational Expectations, Risk and Efficiency in Energy Futures Markets	Chapter 2 uses Fama's (1984) regression approach to measure the information in crude oil, heating oil, and unleaded gas futures prices about future spot prices and time varying premiums.
3	Maturity Effects in Energy Futures	It examines the effects of maturity on future price volatility and trading volume. It provides support for the maturity effect hypothesis.
4	Business Cycles and the Behavior of Energy Prices	It tests the theory of storage in crude oil, heating oil, and unleaded gas markets, using the Fama and French (1988) indirect test. It shows that the theory of storage holds for energy markets.

Chapter Number	Chapter Title	Contents
5	A Cointegration Analysis of Petroleum Futures Prices	It uses Johansen's (1988) maximum likelihood approach to estimating long-run relations in multivariate vector autoregressive models and tests for the number of common stochastic trends in a system of crude oil, heating oil, and unleaded gas futures prices.

Chapter 1:

This chapter examines the empirical evidence for random walk type behavior in energy futures prices. In doing so, tests for unit roots in the univariate time-series representation of the daily crude oil, heating oil, and unleaded gasoline series are performed using recent state-of-the-art methodology. The results show that the unit root hypothesis can be rejected if allowance is made for the possibility of a one-time break in the intercept and the slope of the trend function at an unknown point in time.

Chapter 2:

Conditional on the hypothesis that energy markets are efficient or rational, this chapter uses Fama's (1984) regression approach to measure the information in energy futures prices about future spot prices and time varying premiums. It finds that the premium and expected future spot price components of energy futures prices are negatively correlated and that most of the variation in futures prices is variation in expected premiums.

Chapter 3:

This chapter examines the effects of maturity on future price volatility and trading volume for 129 energy futures contracts recently traded in the NYMEX. The results provide support for the maturity effect hypothesis — that is, energy futures prices do become more volatile and trading volume increases as futures contracts approach maturity.

Chapter 4:

Chapter 4 tests the theory of storage — the hypothesis that the marginal convenience yield on inventory falls at a decreasing rate as inventory increases — in energy markets (crude oil, heating oil, and unleaded gas markets). It uses the Fama and French (1988) indirect test, based on the relative variation in spot and futures prices. The results suggest that the theory holds for the energy markets.

Chapter 5:

This chapter presents evidence concerning the number of common stochastic trends in a system of three petroleum futures prices (crude oil, heating oil, and unleaded gasoline) using daily data from December 3, 1984 to April 30, 1993. Johansen's (1988) maximum likelihood approach for estimating long-run relations in multivariate vector autoregressive models is used. The results indicate the presence of only one common trend.

Chapter 1

Unit Root Behavior in Energy Futures Prices

Apostolos Serletis[*]

1.1   Introduction

Recently the efficient markets hypothesis and the notions connected with it have provided the basis for a great deal of research in financial economics. A voluminous literature has developed supporting this hypothesis. Briefly stated, the hypothesis claims that asset prices are rationally related to economic realities and always incorporate all the information available to the market. This implies that price changes should be serially random, and hence the absence of exploitable excess profit opportunities.

Despite the widespread allegiance to the notion of market efficiency, a number of studies have suggested that certain asset prices are not rationally related to economic realities. For example, Summers (1986) argues that market valuations differ substantially and persistently from rational valuations and that existing evidence (based on common techniques) does not establish that financial markets are efficient.

Market efficiency requires that price changes are uncorrelated and implies a unit root in the level of the price or logarithm of the price series. This is consistent with the empirical work of Nelson and Plosser (1982) who argue that most macroeconomic time series have a unit root (a stochastic trend). Nelson and Plosser described this property as one of being

*Originally published in *The Energy Journal* 13 (1992), 119-128. Reprinted with permission.

"difference stationary" (DS) so that the first difference of a time series is stationary. An alternative "trend stationary" (TS) model, where a stationary component is added to a deterministic trend term, has generally been found to be less appropriate.

Perron (1989), however, challenged this view and argued that most macroeconomic time series (and in particular those used by Nelson and Plosser) are TS if one allows for structural changes in the trend function. In particular, Perron's argument is that only certain "big shocks" have had permanent effects on the various macroeconomic time series and that these shocks were exogenous — that is, not a realization of the underlying data generation mechanism of the various series. Modelling such shocks as exogenous removes the influence of these shocks from the noise function and, in general, leads to a rejection of the null hypothesis of a unit root.

Given the serious implications of unit roots for both empirical and theoretical work as well as the stakes in this line of research, this chapter examines the empirical evidence for random walk type behavior in energy futures prices. The remainder of the chapter consists of three sections. Section 1.1 briefly describes the data. Section 1.2 considers alternative tests of the unit root null hypothesis and presents the results. The final section summarizes the chapter.

1.2   Data

To examine the empirical evidence for random walk type behavior in energy futures prices, I use daily observations from the New York Mercantile Exchange (NYMEX) on spot-month futures prices for crude oil, heating oil and unleaded gasoline. The sample period is 83/07/01 to 90/07/03 for all commodities except unleaded gasoline, which begins in 85/03/14. Figures 1.1 to 1.3 graph the (logarithm of the) price series.

1.3   Empirical Evidence

1.3.1   Autocorrelation Based Tests

Since market efficiency requires serial independence of returns, serial correlation coefficients of orders one to ten have been computed for each contract series and are presented in Table 1.1. Panel A of Table 1.1 contains autocorrelations of the log contract prices. These autocorrelations suggest that (log) contract prices are highly autocorrelated. In particular, the first-order autocorrelations are greater than .993 for every series and the smallest of the autocorrelation coefficients is .920.

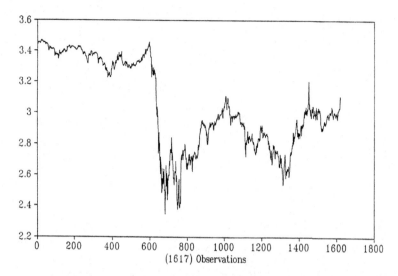

Figure 1.1: Crude Oil Daily 1-Month Log Futures Prices:
01/07/83-03/07/90

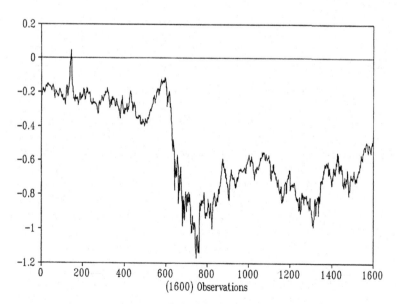

Figure 1.2: Heating Oil Daily 1-Month Log Futures Prices:
01/07/83-03/07/90

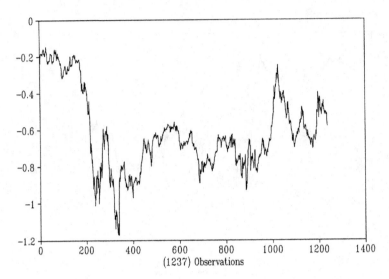

Figure 1.3: Unleaded Gas Daily 1-Month Log Futures Prices:
01/07/83-03/07/90

TABLE 1.1
SAMPLE AUTOCORRELATIONS OF DAILY DATA

Series	r_1	r_2	r_3	r_4	r_5	r_6	r_7	r_8	r_9	r_{10}
A. Logarithms of Daily Contract Prices										
Crude oil	.996	.993	.989	.986	.982	.980	.977	.973	.970	.968
Heating oil	.996	.992	.987	.984	.980	.976	.973	.970	.967	.964
Unleaded gas	.993	.985	.976	.968	.960	.952	.943	.935	.927	.920
B. First Differences of Logarithms of Daily Contract Prices										
Crude oil	-.024	-.049	-.010	.061	-.125	.026	.041	-.058	-.035	.033
Heating oil	.004	-.008	-.025	-.037	-.006	-.087	.069	-.097	-.004	-.004
Unleaded gas	.062	.023	-.029	.033	-.045	.013	.040	-.098	-.018	.084

Note: The sample period is 83/07/01 to 90/07/03 for all commodities except
for unleaded gasoline, which begins in 85/03/14.

Panel B of Table 1.1 reports results in the same fashion as panel A, except that now the first differences of the log contract prices (which measure contract returns) are being considered. Clearly, contract returns are not autocorrelated, suggesting that the hypothesis of (weak form) efficiency cannot be rejected — that is, the past history of returns offers no opportunities for extraordinary profits.

1.3.2 Univariate Tests for Unit Roots

It was argued earlier that market efficiency implies a unit root in the level of the price or logarithm of the price series. Here, using the Philips and Perron (1988) procedure, I test whether the univariate processes of the (natural) logarithms of spot-month energy futures prices contain unit roots. This is a general approach and exploits recent developments in functional central limit theory in order to obtain nonparametric corrections for infinite-dimensional nuisance parameters. The basic idea is to estimate one of two non-augmented Dickey-Fuller regressions defined from

$$y_t = \mu^* + \alpha^* y_{t-1} + u_t^* \tag{1.1}$$

$$y_t = \tilde{\mu} + \tilde{\beta}(t - T/2) + \tilde{\alpha} y_{t-1} + \tilde{u}_t \tag{1.2}$$

where T denotes the sample size.

Given equation (1.1), the null hypotheses of a unit root, with or without a drift, i.e. $H_0^1 : \alpha^* = 1$ and $H_0^2 : \mu^* = 0$, $\alpha^* = 1$, are tested against the stationary alternatives by means of the adjusted t- and F-statistics $Z(t_\alpha^*)$ and $Z(\phi_1)$. In equation (1.2), which allows for a deterministic trend, the null hypotheses $H_0^3 : \tilde{\alpha} = 1$, $H_0^4 : \tilde{\beta} = 0$, $\tilde{\alpha} = 1$, and $H_0^5 : \tilde{\mu} = 0$, $\tilde{\beta} = 0$, $\tilde{\alpha} = 1$ can be tested by means of the test statistics $Z(t_{\tilde{\alpha}})$, $Z(\phi_3)$, and $Z(\phi_2)$, respectively. The formulae for the Z statistics are not presented here. They are derived in Perron (1990) and discussed and applied in Perron (1988).

The results of applying the Z statistics are presented in Table 1.2. The simple unit root test of the t-statistic type, $Z(t_\alpha^*)$, as well as the $Z(\phi_1)$ statistic, are insignificant (at the 1% level) for all the series. The inclusion of a time trend as in (1.2) and the use of the $Z(t_{\tilde{\alpha}})$, $Z(\phi_3)$, and $Z(\phi_2)$ statistics do not change the qualitative results. The overall conclusion is that the evidence is (reasonably) supportive of the unit root hypothesis.

TABLE 1.2

Tests for Unit Roots in the Logarithms of Daily Data

Commodity	$Z(t_{\hat{\alpha}}^*)$	$Z(\phi_1)$	$Z(t_{\hat{\alpha}})$	$Z(\phi_3)$	$Z(\phi_2)$
Crude oil	-1.786	2.204	-1.949	2.274	1.517
Heating oil	-1.817	2.269	-2.194	2.484	1.658
Unleaded gas	-2.412	3.362	-2.059	3.386	2.258

Note: (i) Significant at the **1%, *5%, and +10% level. See Fuller (1976, Table 8.5.2) and Dickey and Fuller (1981), Table IV) for the critical values.

Perron (1989), however, argues that most macroeconomic time series (and in particular those used by Nelson and Plosser, 1982) are trend stationary if one allows for a one-time change in the intercept or in the slope (or both) of the trend function. The postulate is that certain shocks (such as, in the present case, the November 1985 regime shift in OPEC pricing policy as well as the precipitous decline of crude oil prices from \$31 per barrel to the \$9 level, beginning in late 1985) do not represent a realization of the underlying data-generating mechanism of the series under consideration and that the null should be tested against the trend-stationary alternative by allowing, under both the null and alternative hypotheses, for the presence of a one-time break (at a known point in time) in the intercept or in the slope (or both) of the trend function.

Perron's (1989) assumption that the break point is uncorrelated with the data has been criticized, most notably by Christiano (1988) who argues that problems associated with "pre-testing" are applicable to Perron's methodology and that the structural break should instead be treated as being correlated with the data. More recently, Zivot and Andrews (1992), in the spirit of Christiano (1988), treat the selection of the break point as the outcome of an estimation procedure and transform Perron's (1989) conditional (on structural change at a known point in time) unit root test into an unconditional unit root test.

Following Zivot and Andrews (1992), I test the null hypothesis of an integrated process with drift against the alternative hypothesis of trend stationarity with a eon-time break in the intercept and slope of the trend function at an unknown point in time, using the following augmented regression equation (see Zivot and Andrews for more details):

$$y_t = \hat{\mu} + \hat{\theta} DU_t(\hat{\lambda}) + \hat{\beta} t + \hat{\gamma} DT_t(\hat{\lambda}) + \hat{\alpha} y_{t-1} + \sum_{i=1}^{k} \hat{c}_i \Delta y_{t-i} + \hat{e}_i. \quad (1.3)$$

In equation (1.3), testing the null hypothesis of a unit root amounts to choosing the break fraction λ — the ratio of pre-break sample size to total sample size — in order to minimize the one-sided t-statistic for testing $\alpha = 1$. In particular, I reject the null hypothesis of a unit root if $t_{\hat{\alpha}}(\hat{\lambda}) < t(\hat{\lambda})$ where $t(\hat{\lambda})$ denotes the "estimated break point" critical value reported in Zivot and Andrews.

Table 1.3 presents the results using regression (1.3) with λ chosen so as to minimize the one-sided t-statistic for testing $\alpha = 1$ over all $T - 2$ regressions (where T is the number of observations). For each tentative choice of λ, I chose the truncation lag parameter, k, to be correlated with the data. In particular, working backwards from $k = 15$, I chose k such that the t-statistic on the last included lag in the autoregression was greater

TABLE 1.3
TESTS FOR A UNIT ROOT USING ZIVOT AND ANDREWS' PROCEDURE

Regression: $y_t = \hat{\mu} + \hat{\theta}(DU_t(\hat{\lambda})) + \hat{\beta}t + \hat{\gamma}DT_t(\hat{\lambda}) + \hat{\alpha}y_{t-1} + \sum_{i=1}^{k} \hat{c}_i \Delta y_{t-i} + \hat{e}_t$

Series	T	$\hat{T}_B$	k	$\hat{\mu}$	$\hat{\theta}$	$\hat{\beta}$	$\hat{\gamma}$	$\hat{\alpha}$	$S(\hat{e})$
Crude oil	1604	626	12	.087	-.016	-.000	.000	.974*	.022
				(5.0)	(-4.5)	(-1.0)	(2.6)	(-5.1)	
Heating oil	1587	616	12	-.005	-.017	-.000	.000	.972*	.022
				(-2.7)	(-4.9)	(-0.5)	(2.0)	(-5.3)	
Unleaded gas	1225	211	11	-.004	-.014	-.000	.000	.967*	.022
				(-1.2)	(-3.1)	(-1.4)	(1.8)	(-5.1)	

Note: t-statistics are in parentheses. The t-statistic for $\hat{\alpha}$ is the minimum t-statistic over all $T - 2$ regressions for testing $\alpha = 1$. It was determined by estimating equation (3) with the break point, T_B, ranging from $t = 2$ to $t = T - 1$. The t-statistic for $\hat{\alpha}$ is significant at the **1%, *5%, and +10% level. The asymptotic critical values for $t(\hat{\alpha})$ at the 1%, 5%, and 10% significance levelare — -5.57, -5.08, and -4.82, respectively — see Zivot and Andrews (1990, Table 4A).

than 1.6 in absolute value and that the t-statistic on the last lag in higher order autoregressions was less than 1.6. The t-statistics on the parameters for the following respective hypotheses are also presented (in parentheses): $\mu = 0$, $\theta = 0$, $\beta = 0$, $\gamma = 0$, and $\alpha = 1$.

To evaluate the significance of $t_{\hat{\alpha}}(\hat{\lambda})$, the asymptotic "estimated break point" critical values reported in Zivot and Andrews (1992, Table 4A) are used. Clearly, the null hypothesis of a unit root can be rejected at the 5% significance level. Also, the estimated coefficients on the constant $(\hat{\mu})$, the post-break constant dummy $\left(\hat{\theta}\right)$, and the post-break slope dummy $(\hat{\gamma})$ are highly significant. These results imply that the failure of the Phillips-Perron Z statistics to reject the null hypothesis that energy futures prices have a unit root reflects not the presence of the unit root, but instead that the data are trend-stationary about a broken trend.

1.4   Conclusions

This chapter tests for unit roots in the univariate time-series representation of the daily crude oil, heating oil, and unleaded gasoline spot-month futures prices. The results show that the random walk hypothesis for daily energy futures prices can be rejected if allowance is made for the possibility of a one-time break in the intercept and the slope of the trend function at an unknown point in time.

The rejection of the random walk model does not necessarily imply that energy futures markets are inefficient or that energy futures prices are not rational assessments of fundamental values. However, the results highlight the important role that certain big trend breaks could play in tests for unit roots and raise the important question of whether such trend breaks should be treated like any other, or differently, before we classify energy futures prices as either TS or DS.

In addition to its economic importance, the issue of whether energy futures prices are TS or DS has implications for both estimation and hypothesis testing, both of which rely on asymptotic distribution theory. It has been recognized, for example, that inappropriate de-trending of integrated process produces spurious variation in the de-trended series at low frequencies, while inappropriate differencing of trending processes produces spurious variation in the differenced series at high frequencies.

Chapter 2

Rational Expectations, Risk, and Efficiency in Energy Futures Markets

Apostolos Serletis[*]

2.1 Introduction

It is often argued that there are two important social functions of commodity futures markets. First, the transfer of commodity price risk, and, second, the provision of unbiased forecasting by the futures price of the future spot price. Although there is a general consensus that futures markets transfer price risk, there is some debate about the market's forecasting ability. In particular, forecasts based on current spot prices are often as reliable as those based on futures prices.

Serletis and Banack (1990), using recent developments in the theory of cointegration by Engle and Granger (1987), apply efficiency tests to futures and spot energy prices dealing explicitly with the non-stationary nature of those variables. In particular, they test the hypothesis that the futures price is an unbiased predictor of the future spot price and they find evidence consistent with market efficiency. Moreover, they show that the current spot price dominates the current futures price in explaining movements in the future spot price.

---

[*]Originally published in *Energy Economics* (1991), 111-115. Reprinted with permission.

This chapter, conditional on the hypothesis that energy futures markets are efficient or rational, uses Fama's (1984) interesting variance decomposition approach to test a model for joint measurement of variation in the premium and expected future spot price components of energy futures prices. The evidence suggests the presence of a time varying premium. Of course, variation in the premium worsens the performance of the futures price as a predictor of future spot prices.

2.2   Theoretical Foundations

Let $F(t,T)$ be the futures price at time t for delivery of a commodity at T. Let $S(t)$ be the spot price at t. Assuming that the futures price, $F(t,T)$, is the market determined certainty equivalent of the future spot price, $S(T)$, we can split this certainty equivalent into a premium and an expected future spot price (specified in natural logarithms) as

$$F(t,T) = P(t) + E\{S(T)\} \qquad (2.1)$$

where $E\{S(T)\}$ is the rational forecast, conditional on all information available at t, and $P(t)$ is the bias of the futures price, $F(t,T)$, as a forecast of the future spot price, $S(T)$.

Subtracting from both sides of equation (2.1) the current spot price, $S(t)$, we obtain

$$F(t,T) - S(t)P(t) + E\{S(T) - S(t)\} \qquad (2.2)$$

where $F(t,T) - S(t)$, the current futures spot differential, is called the basis. Equation (2.2) implies that the basis can be split into a premium component, $P(t)$, and an expected change in the spot price component. $E\{S(T) - S(t)\}$.

In order to investigate the variability of risk premiums and expected spot price changes as well as their covariability, we use Fama's (1984) simple model for these measurements. In particular, we consider the two complementary regressions of $F(t,T) - S(T)$ and $S(T) - S(t)$ (both observed at T) on $F(t,T) - S(t)$ (observed at t),

$$F(t,T) - S(T) = \alpha_1 + \beta_1 [F(t,T) - S(t)] + u(t,T) \qquad (2.3)$$
$$S(T) - S(t) = \alpha_2 + \beta_2 [F(t,T) - S(t)] + \varepsilon(t,T). \qquad (2.4)$$

Since $F(t,T) - S(T)$ is the premium $P(t)$ plus the random error of the rational forecast, $E\{S(T)\} - S(T)$, estimates of equation (2.3) tell us whether the premium component of the basis has variation that shows up reliably in $F(t,T) - S(T)$. In particular, evidence that β_1 is reliably non-zero means that the basis observed at t, $F(t,T) - S(t)$, contains information about the premium to be realized at T, $F(t,T) - S(T)$. Similarly, estimates of equation (2.4) tell us whether the basis observed at t, has power to predict the future change in the spot price, $S(T) - S(t)$. In fact, evidence that β_2 is reliably non-zero means that the futures price observed at t, $F(t,T)$, has power to forecast the future spot price $S(T)$. Furthermore, the deviation of β_2 from one is a direct measure of the variation of the premium in the futures price.

Equations (2.3) and (2.4) are clearly dependent since the stochastic regressor is the same in both equations and the sum of the dependent variables is the stochastic regressor. The complete complementarity of regressions (2.3) and (2.4) implies that $\hat{\alpha}_1 = -\hat{\alpha}_2$, that $\hat{\beta}_1 = 1 - \hat{\beta}_2$, and that $\hat{u}t,T) = -\hat{\varepsilon}(t,T)$. In other words, regressions (2.3) and (2.4) contain identical information about the variation of the premium and expected change in the spot price components of the basis, and in principle there is no need to estimate both regressions.

Although regressions (2.3) and (2.4) allocate all basis variation to premiums, expected spot-price changes, or some mix of the two, the allocation may be statistically unreliable when the premium and the expected change in the spot price components of the basis are correlated. We can get a good idea about why the regressions may fail to identify the source of variation in the basis by examining the variance of the basis relative to the variance of the premium and expected change in the spot price as well as the covariance between premium and expected change in the spot price.

Following Fama (1984), under appropriate regularity conditions, the probability limits of $\hat{\beta}_1$ and $\hat{\beta}_2$ are given by

$$\beta_1 = \frac{COV[F(t,T) - S(T), F(t,T) - S(t)]}{VAR[F(t,T) - S(t)]} \tag{2.5}$$

$$\beta_2 = \frac{COV[S(T) - S(t), F(t,T) - S(t)]}{VAR[F(t,T) - S(t)]} \tag{2.6}$$

where $COV(.,.)$ and $VAR(.)$ denote the unconditional covariance and variance, respectively. Combining the rational expectations assumption with the decomposition in equation (2.2) implies that

$$\beta_1 = \frac{VAR[P(t)] + COV[P(t), E\{S(T) - S(t)\}]}{VAR[P(t) + VAR[E\{S(T) - S(t)\}] + 2COV[P(t), E\{S(T) - S(t)\}]} \tag{2.7}$$

and

$$\beta_2 = \frac{VAR[E\{S(T) - S(t)\}] + COV[P(t), E\{S(T) - S(t)\}]}{VAR[P(t)] + VAR[E\{S(T) - S(t)\}] + 2COV[P(t), E\{S(T) - S(t)\}]}. \tag{2.8}$$

The importance of equations (2.7) and (2.8) is that when $P(t)$ is constant over time (not necessarily zero), β_1 and β_2 must be identically equal to zero and unity respectively. Hence, the coefficients of β_1 and β_2 describe roughly the degree of variability in the components of the basis. However, only if the premium, $P(t)$, and the expected change in the spot price, $E\{S(T) - S(t)\}$, are uncorrelated would β_1 be equal to the proportion of the variance of the basis due to variance of the risk premium, and under this condition β_2 would be equal to the proportion of the variance of the basis due to variance of the expected change in the spot price. Since it is unlikely that the two components of the basis are uncorrelated, the covariance terms in equations (2.7) and (2.8) must be taken into account. Hence, the simple interpretation of β_1 and β_2 obtained when $P(t)$ and $E\{S(T) - S(t)\}$ are uncorrelated is lost.

2.3 Data

The data include daily observations from the New York Mercantile Exchange (NYMEX) on spot-month and second-month futures prices for heating oil, unleaded gasoline and crude oil. In other words, the spot-month futures prices are used as a proxy for current cash prices, and the second-month futures prices as the current futures prices. The sample period is 1 July 1983 to 31 August 1988 for all commodities except unleaded gasoline, which begins on 14 March 1985.

Under the assumption that the futures price converges to (and therefore predicts) the spot price on the date of the settlement of the futures contract, each current spot and futures price was matched exactly with the spot price on the settlement date of the futures contract. This procedure generated 62 observations for each of heating oil and crude oil and 42 observations for unleaded gasoline.

Table 2.1 shows the standard deviations for the basis, $F(t, T) - S(t)$, the premium, $F(t, T) - S(T)$, and the change in the spot price, $S(T) - S(t)$, for each commodity. For all three commodities, basis variation is low relative to the variation of premiums and spot-price changes, indicating that it is unlikely that regressions (2.3) and (2.4) will reliably assign basis variation to premiums and expected spot price changes.

TABLE 2.1
MEANS AND STANDARD DEVIATIONS OF
$F(t,T) - S(t)$, $F(t,T) - S(T)$ AND $S(T) - S(t)$

Commodity	Observations	Mean	Standard Deviation
Heating Oil	62		
$F(t,T) - S(t)$		-.006	.029
$F(t,T) - S(T)$		-.014	.089
$S(T) - S(t)$		.007	.086
Unleaded Gas	42		
$F(t,T) - S(t)$		-.011	.024
$F(t,T) - S(T)$		-.015	.113
$S(T) - S(t)$		.003	.109
Crude Oil	62		
$F(t,T) - S(t)$		-.007	.012
$F(t,T) - S(T)$		-.001	.098
$S(T) - S(t)$		-.005	.096

Note: $S(t)$, $F(t,T)$ and $S(T)$ are specified in natural logarithms.

My maintained hypothesis is that $F(t,T) - S(t)$, $F(t,T) - S(T)$ and $S(T) - S(t)$ are stationary processes. In fact the hypothesis of univariate stochastic trends is tested following Dickey and Fuller (1979). The null hypothesis for their test (generally called the augmented Dickey-Fuller (ADF) test), is that a series z_t has a unit autoregressive root (i.e. has a stochastic trend). The test is obtained as the t statistic for ρ in the following OLS regression $[\Delta = (1 - L)]$:

$$\Delta z_t = \rho z_{t-1} + \sum_{i=1}^{r} \beta_i \Delta z_{t-i} + \varepsilon_t \qquad (2.9)$$

where z_t is the series under consideration and r is selected to be large enough to ensure that ε_t is a white-noise series. The null hypothesis of stochastically trending z_t is rejected if ρ is negative and significantly different from zero.

In practice, the appropriate order of the autoregression, r, is rarely known. One approach would be to use a model selection procedure based on some information criterion. However, Said and Dickey (1984) showed that the ADF test is valid asymptotically if r is increased with sample size

(N) at a controlled rate, $N^{1/3}$. For my sample sizes, this translates into $r = 4$. It is to be noted that for $r = 0$ the ADF reduces to the simple Dickey-Fuller (DF) test. Also, the distribution of the t test for ρ in equation (2.9) is not standard; rather, it is that given by Fuller (1976).

The DF and ADF tests for stationarity are reported in Table 2.2 for all three commodities. The statistics suggest that all the variables appear to be stationary, i.e. integrated of order one, or $I(1)$ in the terminology of Engle and Granger (1987). Hence standard inference procedures will be applied in the following section.

TABLE 2.2
TESTS FOR UNIT ROOTS IN
$F(t,T) - S(t)$, $F(t,T) - S(T)$ AND $S(T) - S(t)$

Commodity	DF	ADF
Heating Oil		
$F(t,T) - S(t)$	-3.976	-4.073
$F(t,T) - S(T)$	-6.903	-3.970
$S(T) - S(t)$	-8.392	-4.232
Unleaded Gas		
$F(t,T) - S(t)$	-1.820	-1.859
$F(t,T) - S(T)$	-5.581	-3.316
$S(T) - S(t)$	-6.782	-3.399
Crude Oil		
$F(t,T) - S(t)$	-2.870	-1.938
$F(t,T) - S(T)$	-6.920	-3.600
$S(T) - S(t)$	-7.013	-3.325

Note: The asymptotic critical values of DF and ADF at the 1%, 5% and 10% levels are [for 50 observations] -2.62, -1.95, and -1.61, respectively — see Fuller (1976, Table 8.5.2).

2.4   Regression and Cointegration Tests

Table 2.3 shows the estimated regressions of $F(t,T) - S(T)$ and $S(T) - S(t)$ on $F(t,T) - S(t)$. Because of the complementarity of the premium and change regressions, only one set of coefficient standard errors is shown, although the intercepts of α_1 and α_2 and the slopes of β_1 and β_2 for both equation (2.3) and (2.4) are reported. Note that the sum of the intercepts is zero and the sum of the slopes is one. Also, the coefficients of determination R_1^2 and R_2^2 — for the premium and change regressions, respectively — are small since the regressor $F(t,T) - S(t)$ has low variation relative to both $F(t,T) - S(T)$ and $S(T) - S(t)$, as it was also documented in Table 2.1. Moreover, the hypothesis that $\beta_2 = 1$ (or equivalently that $\beta_1 = 0$) is rejected — thus suggesting the presence of a time varying premium.

Turning to the coefficient estimates, the strange numbers in Table 2.3 are the estimates of the regression slope coefficients, $\hat{\beta}_1$ and $\hat{\beta}_2$ for unleaded gas and crude oil. As it was explained earlier, β_1 contains the proportion of the variance of the basis due to variation in its premium component while β_2 contains the proportion of the variance of the basis due to variation in its expected change in the spot price component. These coefficients, however, cannot be interpreted along these lines since the slope coefficients in the premium regressions are almost always greater than one so that those in the change regressions are negative.

For heating oil, however, the evidence that β_1 and β_2 are both positive and less than 1.0 implies reliably positive variances for the premium and the expected change in the spot price. In other words, the futures price has reliable power to forecast spot prices and the futures price contains a time varying premium that shows up reliably in $F(t,T) - S(T)$.

We can get an explanation for the strange estimates of β_1 and β_2 for unleaded gas and crude oil by considering the explicit interpretation of the regression slope coefficients provided by equations (2.7) and (2.8). Inspection of equations (2.7) and (2.8) indicates that since $VAR[F(t,T)-S(t)]$ must be non-negative, a negative estimate of β_2 implies that $COV[P(t), E\{S(T) - S(t)\}]$ is negative and larger in magnitude than $VAR[E\{S(T) - S(t)\}]$. The complementary estimate of $\beta_1 > 1$ then implies that $COV[P(t), E\{S(T)-S(t)\}]$ is smaller in absolute magnitude than $VAR[P(t)]$. Hence, we can conclude that both the premium, $P(t)$, and the expected change in the spot price $E\{S(t) - S(t)\}$ vary through time, and that $VAR[E\{S(T) - S(t)\}]$ is smaller than $VAR[P(t)]$.

In short, except for crude oil, the negative covariation between $P(t)$ and $E\{S(T) - S(t)\}$ prevents us from using the regression coefficients to estimate the levels of $VAR[P(t)]$ and $VAR[E\{S(T) - S(t)\}]$. We can estimate, however, the difference between the two variances as a proportion of $VAR[F(t,T) - S(t)]$

$$\beta_1 - \beta_2 = \frac{VAR[P(t)] - VAR[E\{S(T) - S(t)\}]}{VAR[F(t,T) - S(t)]}. \qquad (2.10)$$

The differences between $\hat{\beta}_1$ and $\hat{\beta}_2$ in Table 2.3 range from 1.160 (unleaded gas) to 1.608 (crude oil) suggesting that, except for heating oil, the difference between the variance of the premium and the variance of the expected change in the spot price is more than the variance of the basis.

TABLE 2.3

REGRESSIONS OF THE PREMIUM $F(t,T) - S(T)$, AND THE CHANGE IN THE SPOT PRICE, $S(T) - S(t)$, ON THE BASIS, $F(t,T) - S(t)$:

$$F(t,T) - S(T) = \hat{\alpha}_1 + \hat{\beta}_1 \{F(t,T) - S(t)\} + \hat{u}(t,T),$$

$$S(T) - S(t) = \hat{\alpha}_2 + \hat{\beta}_2 \{F(t,T) - S(t)\} + \hat{\varepsilon}(t,T)$$

Commodity	$\hat{\alpha}_1$	$\hat{\beta}_1$	$\hat{\alpha}_2$	$\hat{\beta}_2$	$S(\hat{\alpha})$	$S(\hat{\beta})$	R_1^2	R_2^2	DW
Heating Oil	-.009	.785	.009	.215	.011	.383	.065	.005	2.135
Unleaded Gas	-.002	1.080	.002	-.080	.018	.701	.055	.001	2.167
Crude Oil	.008	1.304	-.008	-.304	.014	.995	.027	.001	1.827

Note: R_1^2 and R_2^2 are the coefficients of determination (regression R^2) for the premium and change regressions, respectively. The complete complementarity of the premium and change regressions for each commodity means that the standard errors $S(\hat{\alpha})$ and $S(\hat{\beta})$ of the estimated regression coefficients are the same for the two regressions.

2.5   Conclusion

Regressions of $F(t,T) - S(T)$ on $F(t,T) - S(t)$ are used to test for information in energy futures prices about variation in premiums. Similarly, regressions of $S(T) - S(t)$ on $F(t,T) - S(t)$ are used to test for information in futures prices about future spot prices. The evidence shows that there is variation in both $P(t)$ and $E\{S(T) - S(t)\}$ components of $F(t,T) - S(t)$ and that the variance of the premium component of $F(t,T) - S(t)$ is larger relative to the variance of the expected change in the spot price. Variation in the premium worsens the performance of the futures price as a predictor of future spot prices.

Chapter 3

Maturity Effects in Energy Futures

*Apostolos Serletis**

3.1 Introduction

The price variability of futures contracts has attracted a great deal of attention and has been explored extensively in the literature during the past two decades, ever since Samuelson (1965) advanced the hypothesis that (under the assumption that spot prices follow a stationary first-order autoregressive process, and futures prices are unbiased estimates of the settlement spot price) the variance of futures prices increases as the futures contract approaches maturity. It has been argued that there is strong empirical support for a maturity effect in volatility although there appear to be differences regarding the importance of the maturity effect.

Knowledge of futures price variability is, in general, essential to the margin-setting authority. In particular, the minimum margins (this is, down-payments) that futures brokers require of futures traders depend upon the price variability of futures contracts. Indeed, the higher the futures price variability, the higher the minimum margins (presumably to reduce speculation and volatility).

As mentioned previously, many researchers have studied the relationship between maturity and the volatility of futures prices over the life of a large

*Originally published in *Energy Economics* (1992), 150-157. Reprinted with permission.

number of agricultural and financial contracts. A quick survey of the literature reveals a certain consensus. It is only regarding the importance of the
maturity effect (relative to other factors driving price volatility) that there
appear to be differences of opinion. For example, Fama and French (1988)
along with the earlier evidence by Anderson (1985) and Milonas (1986) all
suggest that the maturity effect exists in commodity prices. Other studies test for the persistence of a maturity effect after the volume of trade
is introduced as additional explanatory variable. Thus Grammatikos and
Saunders (1986) fail to find a maturity effect when controlling for the volume of trade on volatility. Nevertheless, it is still interesting to examine
the maturity effect on recent data, or different commodities and/or using
new methodologies.

Motivated by these considerations, this chapter examines the effect of
maturity on energy futures price variability and trading volume using a
method for measuring the variability of futures prices proposed by Parkinson (1980) and Garman and Klass (1980). In doing so, this chapter utilizes
daily high and low prices and daily trading volume for 129 energy futures
contracts recently traded in the New York Mercantile Exchange (NYMEX).
The empirical evidence supports the hypothesis that energy futures prices
become more volatile and trading volume increases as futures contracts near
maturity.

3.2   Data and the Measurement of Futures Price Variability

The data consist of daily high and low prices and daily trading volume for
43 futures contracts in three different energy futures (i.e., 129 contracts
in all) traded in the New York Mercantile Exchange (NYMEX): crude oil,
heating oil and unleaded gas. For each contract, the daily high and how
price and volume were traced from the inception of the contact to its expiry.
The maturity dates range from January 1987 to July 1990.

The method of measuring price variability, following Parkinson (1980)
and Garman and Klass (1980), takes advantage of all readily available information in contrast to the classical approach which employs only the variance
of the daily logarithmic price changes — that is $VAR[\ln P(t) - \ln P(t-1)]$,
where $P(t)$ is the closing price on day t. Specifically, assuming that prices
follow a random walk with zero drift then a metric of price variability on
day t is given by:

$$VAR(t) = \frac{[\ln H(t) - \ln L(t)]^2}{4\ln 2} \qquad (3.1)$$

where $H(t)$ and $L(t)$ are, respectively, the high and low prices on day t.

In order to examine the effect of maturity on price variability, the following regression equations were estimated using ordinary least squares (OLS):

$$VAR(t) = \alpha_0 + \alpha_1 \ln t + \varepsilon(t) \tag{3.2}$$

$$VAR(t) = \beta_0 + \beta_1 \ln t + \beta_2 \ln VOL(t) + u(t) \tag{3.3}$$

where $VAR(t)$ is given by equation (3.1), t is the number of days remaining until the futures contract expires, $VOL(t)$ is the number of futures contracts traded on day t, and the error terms, $\varepsilon(t)$ and $u(t)$ are each assumed to be independently and identically distributed.

The constant term α_0 measures the price variability at maturity and should be positive under the assumption that it equals the variability of the spot price on that day. The slope coefficient α_1 measures the sensitivity of the variability of the futures price to changes in time to maturity. If futures prices do become more volatile (the Samuelson hypothesis), then α_1 in equation (3.2) will be negative and significantly different from zero. The additional regression of price variability on time to maturity and volume [equation (3.3)] is used to check for the presence of a maturity effect in price volatility caused by other factors than those affecting the volume of trade — see, for example, Grammatikos and Saunders (1986).

3.3 Empirical Results

Before proceeding to estimate the models outlined above, there is one further issue that needs to be addressed, and that is the time series properties of the variables involved. Following the analysis of Engle and Granger (1989) and the recent growth in the theory of integrated variables, if the variables are integrated of order one [or $I(1)$] in their terminology], but do not cointegrate, ordinary least squares (OLS) yields misleading results. In fact, Phillips (1987) formally proves that a regression involving integrated variables is spurious in the absence of cointegration. Under these circumstances it becomes important to evaluate empirically the time series properties of the variables involved.

In the first two columns of Tables 3.1-3.3, I report the Dickey-Fuller (DF) test for stationarity of the price volatility and log trading volume for 43 futures contracts in three different energy futures: crude oil, heating oil and unleaded gasoline. The results give a rather unambiguous picture. The DF statistic suggests that price volatility and trading volume are stationary quantities and that traditional distribution theory is applicable.

The estimation results of equations (3.2) and (3.3) are reported in columns 3-7 and 8-14, respectively, of Tables 3.1-3.3. The results in columns

3-7 suggest that almost all the contracts exhibit an α_0 significantly positive and that over 70% of the contracts have an α_1 significantly (at the 10% level) negative. These results appear to support the maturity effect hypothesis. Note, however, that the R^2 statistics seem to indicate that, in general, time to maturity explains very little of price variability.

Once the volume of trade is introduced as additional explanatory variable [as in equation (3.3)], the results in columns 8-14 of Tables 3.1-3.3 clearly suggest that only 35 of the contracts have a β_1 significantly (at the 10% level) negative. What this means is that it is probably not maturity per se which affects volatility, but rather one or more factors which simultaneously affect the volume of trade and volatility.

Figures 3.1 to 3.12 graph price variability and trading volume against time to maturity for some representative energy futures contracts. Clearly, trading volume increases initially and then falls off as the contract approaches maturity. Price variability also appears to exhibit such behaviour, however, the "peak" tends to occur closer to maturity than for trading volume. This could be due to the last month of trade "expiration effects" in volatility and trading volume which are primarily caused by hedgers and speculators switching to the next available contract during this month.

So far I have analyzed contemporaneous relations between price volatility and trading volume ignoring any potential lead (lag) relations between these variables. In this section, I test the direction of possible causality between price volatility and trading volume, in the sense of Granger (1969). In particular, I investigate whether knowledge of past trading volume improves the prediction of futures price volatility beyond predictions that are based on past futures price volatility alone. This is the empirical definition of Granger causality.

To test the direction of causality between futures price volatility and trading volume it must be assumed that the relevant information is entirely contained in the present and past values of these variables. A specification that suggests itself is

$$VAR_t = \varphi_0 + \sum_{i=1}^{r} \varphi_i VAR_{t-i} + \sum_{j=1}^{s} \theta_j \ln(VOL)_{t-j} + u_t \qquad (3.4)$$

where u_t is a disturbance term. To test if trading volume causes futures price variability in the Granger sense, we first estimate (3.4) by ordinary least squares to obtain the unrestricted sum of squared residuals, SSR_u. Then, by running another regression under the restriction that all θ_j's are zero, the restricted sum of squared residuals, SSR_r, is obtained. If u_t is white noise, then the statistic computed as the ratio of $(SSR_r - SSR_u)/s$ to $SSR_u/(n-r-s-1)$ has an asymptotic F-distribution with the numerator

TABLE 3.1

THE MATURITY EFFECT ON PRICE VARIABILITY FOR CRUDE OIL

Contract	DF VAR ln(VOL)		System $VAR = \alpha_0 + \alpha_1 \ln t$					$VAR = \beta_0 + \beta_1 \ln t + \beta_2 \ln(VOL)$								
			$\hat{\alpha}_0$	$t(\hat{\alpha}_0)$	$\hat{\alpha}_1$	$t(\hat{\alpha}_1)$	R^2	β_0	$t(\beta_0)$	β_1	$t(\beta_1)$	β_2	$t(\beta_2)$	R^2	F_1	F_2
87/1	-12.7	-3.3	-0.9E-02	0.6	0.3E-02	0.9	0.005	0.8E-02	0.2	0.1E-02	0.2	-0.1E-02	0.5	0.007	0.698	0.691
87/2	-8.5	-2.7	0.4E-04	0.3	0.5E-04	1.5	0.014	-07E-03	2.0	0.1E-03	2.8	0.5E-04	2.4	0.050	0.209	0.083
87/3	-9.6	-3.0	-0.9E-04	0.8	0.7E-04	2.9	0.051	-0.4E-03	1.5	0.1E-03	2.8	0.2E-04	1.3	0.044	0.022*	0.064
87/4	-12.2	-2.8	0.9E-04	1.3	0.1E-04	0.5	0.002	0.1E-04	0.0	0.2E-04	0.7	0.5E-05	0.4	0.003	0.980	0.483
87/5	-11.8	-2.7	0.9E-04	1.3	0.7E-05	0.4	0.001	-0.1E-03	0.7	0.3E-04	1.4	0.1E-04	1.4	0.013	0.248	0.116
87/6	-11.4	-2.8	0.6E-04	1.6	0.3E-05	0.3	0.001	-0.1E-03	1.2	0.2E-04	1.9	0.1E-04	2.2	0.033	0.815	0.363
87/7	-11.8	-2.4	0.1E-03	2.5	-0.1E-04	1.2	0.009	0.1E-03	1.2	-0.1E-04	0.8	-0.1E-06	0.1	0.009	0.887	0.549
87/8	-10.9	-3.0	0.2E-03	4.3	-0.4E-04	3.2	0.063	0.2E-03	2.0	-0.5E-04	2.2	-0.2E-05	0.2	0.063	0.292	0.566
87/9	-11.1	-2.6	0.2E-03	6.3	-0.3E-04	4.4	0.113	0.5E-04	0.7	-0.1E-04	1.3	0.9E-05	0.1	0.135	0.013*	0.913
87/10	-12.8	-2.5	0.1E-02	2.2	-0.2E-03	1.9	0.022	0.1E-02	0.9	-0.2E-03	1.2	-0.9E-06	0.1	0.022	0.203	0.951
87/11	-10.0	-2.7	0.2E-03	6.5	-0.3E-04	4.7	0.121	0.2E-03	0.3	-0.1E-04	1.0	0.1E-04	2.7	0.159	0.004*	0.913
87/12	-2.5	-3.0	0.9E-03	5.3	-0.2E-03	4.8	0.121	0.2E-02	4.8	-0.3E-03	5.5	-0.1E-03	3.0	0.168	0.416	0.370
88/1	-3.2	-2.4	0.7E-03	11.7	-0.1E-03	10.4	0.398	0.1E-02	8.8	-0.2E03	10.4	-0.3E-04	4.2	0.457	0.306	0.852
88/2	-5.9	-2.9	0.6E-03	7.3	-0.1E-03	6.0	0.183	0.3E-05	0.0	-0.4E-04	1.6	0.4E-04	3.2	0.234	0.005*	0.306
88/3	-5.8	-2.9	0.3E-03	4.3	-0.5E-04	2.9	0.049	-0.7E-03	4.1	0.7E-04	2.9	0.7E-04	6.5	0.246	0.032*	0.860
88/4	-6.9	-2.8	0.7E-03	6.9	-0.1E-03	5.7	0.165	0.7E-03	2.6	-0.1E-03	3.6	0.2E-05	0.1	0.165	0.024*	0.852
88/5	-7.0	-3.1	0.2E-03	2.7	-0.2E-04	1.2	0.009	-0.4E-03	2.4	0.5E-04	2.2	0.4E-04	4.2	0.109	0.420	0.794
88/6	-6.4	-3.8	0.1E-03	1.7	-0.6E-06	0.1	0.000	-0.3E-03	2.8	0.5E-04	2.9	0.3E-04	4.2	0.101	0.613	0.370
88/7	-7.1	-4.2	0.1E-03	2.1	-0.1E-05	0.1	0.000	-0.1E-03	1.4	0.3E-04	1.8	0.1E-04	2.6	0.042	0.827	0.001*
88/8	-7.6	-2.9	0.5E-03	8.2	-0.9E-04	6.5	0.208	0.7E-03	4.9	-0.1E-03	5.4	-0.1E-04	1.5	0.219	0.085	0.335
88/9	-10.1	-2.9	0.3E-03	4.9	-0.5E-04	3.4	0.068	-0.2E-03	1.4	0.1E-04	0.6	0.3E-04	4.5	0.176	0.001*	0.887
88/10	-11.4	-2.8	0.5E-03	5.1	-0.9E-04	3.8	0.085	0.1E-04	0.0	-0.2E-04	0.8	0.3E-04	2.5	0.121	0.003*	0.895
88/11	-6.7	-4.4	0.8E-03	12.4	-0.1E-03	10.5	0.430	0.8E-03	5.5	-0.1E-03	7.1	-0.1E-05	4.1	0.403	0.016*	0.550
88/12	-8.0	-2.7	0.6E-03	8.9	-0.1E-03	6.5	0.208	0.1E-04	0.0	-0.3E-04	1.6	0.4E-04	4.1	0.283	0.001*	0.818

TABLE 3.1 CONT'D

Contract	DF VAR ln(VOL)	System $VAR = \alpha_0 + \alpha_1 \ln t$ $\hat{\alpha}_0$	$t(\hat{\alpha}_0)$	$\hat{\alpha}_1$	$t(\hat{\alpha}_1)$	R^2	$VAR = \beta_0 + \beta_1 \ln t + \beta_2 \ln(VOL)$ β_0	$t(\beta_0)$	β_1	$t(\beta_1)$	β_2	$t(\beta_2)$	R^2	F_1	F_2
89/1	-12.4	-3.2	0.2E-02	4.0	-0.6E-03	3.5	0.071	0.2E-02	1.4	-0.5E-03	2.1	0.1E-04	0.1	0.071	0.019*	0.523
89/2	-8.3	-2.8	0.5E-03	4.7	-0.7E-04	2.9	0.051	-0.8E-04	0.3	-0.2E-05	0.1	0.3E-04	2.4	0.086	0.077	0.355
89/3	-10.9	-3.7	0.4E-03	5.2	-0.6E-04	3.3	0.064	-0.2E-03	1.0	0.1E-04	0.5	0.4E-04	3.6	0.135	0.006*	0.818
89/4	-8.8	-3.7	0.3E-03	6.0	-0.5E-04	3.5	0.072	0.2E-05	0.0	-0.5E-05	0.2	0.2E-04	2.6	0.111	0.009*	0.601
89/5	-5.2	-2.5	0.8E-03	7.6	-0.1E-03	6.0	0.183	0.1E-02	4.8	-0.2E-03	5.4	-0.4E-04	2.2	0.208	0.052	0.698
89/6	-9.3	-3.3	0.4E-03	6.7	-0.5E-04	4.0	0.091	-0.5E-05	0.0	-0.1E-04	0.4	0.2E-04	2.6	0.127	0.003*	0.012*
89/7	-8.1	-3.3	0.1E-02	8.7	-0.2E-03	7.2	0.247	0.1E-02	6.0	-0.2E-03	6.9	-0.4E-04	2.6	0.278	0.763	0.370
89/8	-11.2	-2.7	0.4E-03	4.5	-0.7E-04	2.9	0.501	0.1E-05	0.0	-0.2E-04	0.5	0.3E-04	1.7	0.068	0.012*	0.869
89/9	-11.7	-4.2	0.1E-03	3.1	-0.9E-05	0.6	0.002	-0.4E-03	3.0	-0.2E-04	2.9	0.4E-04	4.6	0.117	0.226	0.103
89/10	-11.1	-2.9	0.7E-04	1.7	0.9E-05	0.9	0.005	-0.3E-03	2.6	0.5E-04	3.2	0.3E-04	3.4	0.071	0.756	0.677
89/11	-11.5	-3.9	0.8E-04	1.9	0.3E-05	0.3	0.000	-0.1E-03	1.1	0.2E-04	1.6	0.1E-04	1.8	0.021	0.818	0.205
89/12	-11.5	-2.9	0.8E-05	0.1	0.2E-04	1.9	0.021	-0.1E-03	0.5	0.3E-04	1.6	0.9E-05	0.6	0.024	0.230	0.818
90/1	-2.8	-2.7	0.3E-03	7.9	-0.5E-04	6.0	0.181	0.6E-03	5.1	-0.9E-04	5.9	-0.2E-04	2.8	0.218	0.446	0.990
90/2	-3.7	-4.2	0.8E-03	11.1	-0.1E-03	9.6	0.362	0.1E-02	6.1	-0.2E-03	7.9	-0.3E-04	2.4	0.306	0.070	0.677
90/3	-7.6	-10.7	0.3E-03	9.2	-0.7E-04	7.0	0.230	-0.2E-03	2.2	-0.9E-06	0.1	0.4E-04	5.4	0.347	0.000*	0.503
90/4	-8.1	-3.1	-0.2E-03	7.7	-0.4E-04	5.6	0.159	-0.1E-03	1.6	-0.8E-06	0.1	0.3E-04	4.7	0.261	0.009*	0.249
90/5	-9.0	-3.6	0.1E-02	6.7	-0.2E-03	5.8	0.174	0.1E-02	3.7	-0.2E-03	4.8	-0.3E-04	1.4	0.184	0.077	0.990
90/6	-8.7	-3.2	0.6E-03	7.8	-0.1E-03	6.3	0.195	0.1E-03	0.7	-0.7E-04	2.4	0.3E-04	2.0	0.215	0.002*	0.827
90/7	-11.4	-2.7	0.1E-02	5.4	-0.2E-03	4.5	0.113	0.4E-03	0.6	-0.1E-03	1.8	0.4E-04	0.8	0.117	0.001*	0.860

Notes: DF refers to the Dickey-Fuller (1979) t-statistic for a unit root. Under the null hypothesis of a unit root the 1%, 5% and 10% critical values of the DF statistic are -2.58, -1.95 and -1.62, respectively — see Fuller (1976, Table 8.5.2). t-ratios are all in absolute terms. An asterisk (in the last two columns) indicates significance (i.e. the null hypothesis of no causality would be rejected at the 5% level).

TABLE 3.2

THE MATURITY EFFECT ON PRICE VARIABILITY FOR HEATING OIL

| Contract | DF VAR | ln(VOL) | System $VAR = \alpha_0 + \alpha_1 \ln t$ | | | | | $VAR = \beta_0 + \beta_1 \ln t + \beta_2 \ln(VOL)$ | | | | | | | | |
			$\hat{\alpha}_0$	$t(\hat{\alpha}_0)$	$\hat{\alpha}_1$	$t(\hat{\alpha}_1)$	R^2	$\hat{\beta}_0$	$t(\hat{\beta}_0)$	$\hat{\beta}_1$	$t(\hat{\beta}_1)$	$\hat{\beta}_2$	$t(\hat{\beta}_2)$	R^2	F_1	F_2
87/1	-7.2	-3.2	0.2E-03	2.6	-0.1E-04	0.4	0.000	-0.5E-03	2.4	0.1E-03	2.8	0.6E-04	4.1	0.086	0.670	0.433
87/2	-8.3	-2.9	0.2E-03	2.9	-0.9E-05	0.4	0.001	-0.2E-03	1.4	0.6E-04	2.0	0.3E-04	3.2	0.056	0.303	0.328
87/3	-12.4	-3.8	0.1E-03	2.3	0.6E-05	0.3	0.000	-0.5E-04	0.3	0.3E-04	1.3	0.1E-04	1.3	0.011	0.771	0.705
87/4	-12.0	-3.3	0.2E-03	3.5	-0.1E-04	1.2	0.008	-0.1E-04	0.1	0.1E-04	0.5	0.1E-04	1.8	0.027	0.185	0.748
87/5	-12.8	-3.9	0.1E-03	2.6	-0.6E-05	0.4	0.000	-0.4E-04	0.2	0.2E-04	0.9	0.1E-04	1.6	0.015	0.125	0.057
87/6	-12.3	-3.3	0.1E-03	3.0	-0.1E-04	0.8	0.004	-0.2E-03	2.1	0.4E-04	2.5	0.2E-04	4.3	0.110	0.046*	0.047*
87/7	-12.2	-2.9	0.11260	2.0	-0.2E-01	1.8	0.022	0.18043	1.6	-0.3E-01	1.7	-0.5E-02	0.6	0.025	0.691	0.990
87/8	-10.1	-3.1	0.1E-03	5.8	-0.2E-04	3.5	0.085	0.7E-04	1.1	-0.1E-04	1.1	0.9E-05	2.1	0.117	0.027*	0.335
87/9	-8.3	-3.4	0.2E-03	9.7	-0.5E-04	7.3	0.272	0.2E-03	4.0	-0.4E-04	4.6	0.4E-05	1.0	0.278	0.077	0.818
87/10	-8.2	-3.8	0.1E-03	7.5	-0.3E-04	5.1	0.147	-0.2E-04	0.5	-0.2E-05	0.2	0.1E-04	5.0	0.270	0.007*	0.664
87/11	-7.4	-3.2	0.2E-03	5.7	-0.3E-04	4.2	0.100	0.4E-06	0.0	-0.5E-05	0.4	0.1E-04	3.7	0.173	0.004*	0.528
87/12	-8.4	-3.2	0.1E-03	5.0	-0.2E-04	2.9	0.045	0.1E-03	1.8	0.1E-04	1.3	0.2E-04	4.8	0.156	0.013*	0.651
88/1	-6.1	-3.3	0.5E-03	8.0	-0.9E-04	6.5	0.209	0.3E-03	2.8	-0.8E-04	3.7	0.8E-05	1.0	0.214	0.032*	0.561
88/2	-7.3	-3.1	0.4E-03	8.3	-0.8E-04	6.5	0.202	0.9E-04	0.7	-0.3E-04	1.7	0.2E-04	3.5	0.257	0.000*	0.941
88/3	-6.6	-2.7	0.3E-03	7.1	-0.6E-04	5.1	0.131	-0.5E-04	0.4	-0.2E-05	0.1	0.2E-04	4.1	0.208	0.001*	0.613
88/4	-6.5	-3.1	0.3E-03	5.2	-0.5E-04	3.5	0.067	-0.1E-03	0.8	0.1E-04	0.6	0.3E-04	4.2	0.155	0.016*	0.980
88/5	-5.8	-3.5	0.2E-03	4.1	-0.3E-04	2.4	0.032	-0.1E-03	0.9	0.2E-04	0.9	0.2E-04	3.5	0.096	0.077	0.664
88/6	-6.6	-3.3	0.1E-03	1.9	-0.4E-05	0.2	0.000	-0.2E-03	1.8	0.4E-04	2.1	0.2E-04	3.1	0.055	0.539	0.886
88/7	-6.8	-4.0	0.1E-03	3.2	-0.1E-04	1.5	0.012	-0.9E-04	0.8	0.1E-04	0.9	0.2E-04	2.7	0.053	0.523	0.895
88/8	-8.7	-2.8	0.4E-03	8.5	-0.8E-04	6.7	0.238	0.3E-03	3.5	-0.7E-04	4.1	0.3E-05	0.5	0.240	0.052	0.951
88/9	-7.6	-3.8	0.2E-03	6.2	-0.4E-04	4.3	0.112	-0.8E-04	1.1	0.4E-05	0.3	0.2E-04	5.5	0.262	0.008*	0.455
88/10	-6.6	-3.5	0.4E-03	9.3	-0.8E-04	7.3	0.259	0.1E-03	1.8	-0.4E-04	3.0	0.1E-04	3.3	0.311	0.002*	0.990
88/11	-7.2	-3.2	0.6E-03	11.3	-0.1E-03	9.1	0.334	0.3E-03	3.2	-0.7E-04	4.1	0.1E-04	3.0	0.371	0.001*	0.498
88/12	-8.2	-2.5	0.6E-03	11.7	-0.1E-03	9.2	0.323	0.3E-03	2.7	-0.7E-04	4.2	0.2E-04	3.4	0.366	0.002*	0.589

TABLE 3.2 CONT'D

Contract	DF VAR	DF ln(VOL)	System VAR $= \alpha_0 + \alpha_1 \ln t$ $\hat{\alpha}_0$	$t(\hat{\alpha}_0)$	$\hat{\alpha}_1$	$t(\hat{\alpha}_1)$	R^2	VAR $= \beta_0 + \beta_1 \ln t + \beta_2 \ln(VOL)$ β_0	$t(\beta_0)$	β_1	$t(\beta_1)$	β_2	$t(\beta_2)$	R^2	F_1	F_2
89/1	-8.8	-4.0	0.3E-03	5.8	-0.4E-04	3.3	0.065	-0.2E-03	2.4	0.2E-04	1.5	0.4E-04	6.7	0.268	0.004*	0.523
89/2	-10.3	-3.6	0.4E-03	7.9	-0.7E-04	5.6	0.146	0.1E-03	0.7	-0.3E-04	1.5	0.2E-04	2.8	0.182	0.003*	0.303
89/3	-9.9	-2.8	0.4E-03	9.2	-0.6E-04	6.1	0.172	0.6E-05	0.1	-0.8E-05	0.4	0.2E-04	4.1	0.242	0.000*	0.951
89/4	-12.4	-3.0	0.7E-03	5.9	-0.1E-03	4.5	0.105	0.5E-03	2.0	-0.1E-03	2.4	0.1E-04	0.6	0.107	0.013*	0.843
89/5	-10.9	-3.7	0.5E-03	8.2	-0.7E-04	5.4	0.138	0.1E-03	1.1	-0.3E-04	1.6	0.2E-04	2.8	0.174	0.001*	0.726
89/6	-13.3	-4.9	0.7E-03	3.1	-0.1E-03	2.1	0.024	0.2E-03	0.4	-0.7E-04	0.8	0.4E-04	0.9	0.029	0.047	0.818
89/7	-10.6	-4.4	0.5E-03	7.4	-0.8E-04	5.0	0.126	0.2E-03	0.2	-0.2E-04	0.9	0.3E-04	4.0	0.202	0.000*	0.670
89/8	-8.6	-3.7	0.4E-03	7.4	-0.6E-04	4.9	0.130	-0.1E-03	0.9	0.4E-05	0.2	0.3E-04	5.6	0.274	0.000*	0.625
89/9	-9.2	-3.0	0.2E-03	4.2	-0.2E-04	1.8	0.019	-0.3E-03	4.1	0.5E-05	3.8	0.4E-04	7.7	0.280	0.001*	0.852
89/10	-10.9	-4.0	0.2E-03	5.0	-0.2E-04	2.4	0.036	-0.1E-03	1.1	0.1E-04	1.1	0.2E-04	3.9	0.122	0.026*	0.852
89/11	-11.9	-4.8	0.1E-03	4.5	-0.1E-04	1.9	0.020	-0.6E-04	0.7	0.1E-04	0.8	0.1E-04	2.8	0.067	0.057	0.923
89/12	-12.6	-4.4	0.1E-03	4.6	-0.1E-04	1.7	0.016	-0.4E-04	0.6	0.1E-04	0.9	0.1E-04	3.1	0.066	0.093	0.047
90/1	-9.1	-3.5	0.2E-02	9.7	-0.6E-03	9.1	0.339	0.6E-02	10.4	-0.1E-02	11.5	-0.2E-03	6.4	0.474	0.748	0.990
90/2	-11.9	-3.4	0.1E-02	4.8	-0.3E-03	4.2	0.093	0.1E-02	2.2	-0.3E-03	2.8	-0.7E-06	0.1	0.093	0.039*	0.932
90/3	-8.4	-3.7	0.4E-03	7.9	-0.8E-04	5.9	0.166	-0.2E-04	0.2	-0.1E-04	0.8	0.3E-04	4.8	0.266	0.001*	0.533
90/4	-9.5	-3.9	0.4E-03	7.9	-0.6E-04	5.6	0.154	-0.5E-04	0.5	-0.1E-04	0.7	0.3E-04	4.9	0.259	0.001*	0.249
90/5	-9.7	-4.1	0.5E-03	8.5	-0.9E-04	6.5	0.188	0.2E-03	1.7	-0.5E-04	2.7	0.2E-04	2.3	0.214	0.001*	0.589
90/6	-8.8	-4.2	0.3E-03	6.4	-0.5E-04	4.2	0.092	-0.1E-03	1.3	0.1E-04	0.6	0.3E-04	5.1	0.209	0.005*	0.225
90/7	-9.5	-3.4	0.3E-03	6.2	-0.4E-04	3.9	0.076	-0.1E-03	1.4	0.1E-04	0.6	0.3E-04	4.4	0.167	0.000*	0.712

Notes: See notes to Table 3.1.

TABLE 3.3

THE MATURITY EFFECT ON PRICE VARIABILITY FOR UNLEADED GASOLINE

Contract	DF VAR	DF ln(VOL)	System $VAR=\alpha_0+\alpha_1\ln t$ $\hat{\alpha}_0$	$t(\hat{\alpha}_0)$	$\hat{\alpha}_1$	$t(\hat{\alpha}_1)$	R^2	$VAR=\beta_0+\beta_1\ln t+\beta_2\ln(VOL)$ $\hat{\beta}_0$	$t(\hat{\beta}_0)$	$\hat{\beta}_1$	$t(\hat{\beta}_1)$	$\hat{\beta}_2$	$t(\hat{\beta}_2)$	R^2	F_1	F_2
87/1		-9.4	0.2E-03	2.3	-0.1E-04	0.7	0.005	0.6E-04	0.3	0.6E-05	0.1	0.1E-04	1.0	0.018	0.188	0.802
87/2		-8.6	0.2E-03	4.1	-0.3E-04	1.9	0.039	-0.7E-04	0.6	0.8E-05	0.4	0.3E-04	3.6	0.157	0.009*	0.567
87/3		-8.7	0.2E-03	4.6	-0.3E-04	2.4	0.060	0.2E-04	0.2	-0.7E-05	0.4	0.1E-04	2.4	0.114	0.026*	0.980
87/4		-8.1	0.1E-03	2.9	-0.9E-05	0.7	0.005	-0.1E-03	1.5	0.2E-04	1.3	0.2E-04	3.3	0.109	0.096	0.475
87/5		-9.5	0.1E-03	4.4	-0.2E-04	2.1	0.040	0.7E-06	0.0	-0.8E-06	0.0	0.1E-04	2.3	0.088	0.049*	0.932
87/6		-7.0	0.9E-04	3.9	-0.6E-05	1.1	0.011	-0.1E-03	2.7	0.1E-04	2.4	0.1E-04	5.2	0.214	0.087	0.607
87/7		-10.5	0.4E-03	2.7	-0.8E-04	2.2	0.042	0.5E-03	1.6	-0.9E-04	1.9	-0.9E-05	0.3	0.043	0.518	0.624
87/8		-8.4	0.1E-03	7.3	-0.3E-04	5.0	0.174	0.1E-03	2.6	-0.2E-04	3.0	0.4E-05	1.1	0.184	0.154	0.304
87/9		-8.5	0.2E-03	8.2	-0.4E-04	5.9	0.223	0.1E-03	2.7	-0.3E-04	3.6	0.4E-05	0.8	0.227	0.015*	0.913
87/10		-10.9	0.6E-03	1.5	-0.1E-03	1.1	0.011	-0.1E-03	0.1	-0.3E-04	1.2	0.6E-04	0.9	0.018	0.304	0.960
87/11		-8.0	0.2E-03	6.0	-0.4E-04	4.3	0.134	0.1E-03	1.3	-0.2E-04	1.9	0.1E-04	1.7	0.156	0.008*	0.460
87/12		-11.1	0.1E-01	1.9	-0.2E-02	1.6	0.022	0.2E-01	1.5	-0.3E-02	1.8	-0.9E-03	0.8	0.028	0.741	0.550
88/1		-2.9	0.4E-03	6.3	-0.8E-04	4.9	0.172	0.2E-03	1.7	-0.6E-04	2.8	0.1E-04	1.3	0.184	0.206	0.787
88/2		-3.6	0.4E-03	5.5	-0.7E-04	4.0	0.121	-0.3E-04	0.2	-0.2E-04	0.9	0.3E-04	3.5	0.208	0.036*	0.904
88/3		-3.6	0.1E-03	2.7	-0.2E-04	1.2	0.012	-0.4E-03	3.0	0.4E-04	2.2	0.5E-04	4.9	0.183	0.154	0.734
88/4		-3.2	0.1E-03	2.3	-0.1E-04	0.7	0.005	-0.1E-03	0.7	0.1E-04	0.6	0.2E-04	2.1	0.041	0.771	0.913
88/5		-3.4	0.1E-03	2.4	-0.9E-05	0.6	0.003	-0.5E-04	0.4	0.1E-04	0.5	0.1E-04	1.7	0.029	0.835	0.651
88/6		-3.5	0.6E-04	1.5	0.1E-05	0.1	0.000	-0.4E-04	0.5	0.1E-04	1.0	0.9E-05	1.4	0.017	0.932	0.154
88/7		-4.3	0.1E-03	6.2	-0.2E-04	3.8	0.107	0.2E-03	2.5	-0.3E-04	2.7	-0.1E-05	0.2	0.108	0.250	0.534
88/8		-2.2	0.1E-01	1.4	-0.4E-02	1.2	0.012	0.5E-02	0.1	-0.2E-02	0.4	0.9E-03	0.4	0.014	0.336	0.913
88/9		-2.9	0.4E-03	6.6	-0.8E-04	4.9	0.172	-0.4E-04	0.3	-0.2E-04	1.0	0.3E-04	4.5	0.297	0.009*	0.960
88/10		-3.2	0.7E-03	10.0	-0.1E-03	8.0	0.349	0.4E-03	2.7	-0.1E-03	4.8	0.2E-04	1.6	0.365	0.127	0.951
88/11		-3.4	0.1E-02	6.7	-0.4E-03	5.8	0.216	0.2E-02	3.8	-0.4E-03	4.9	-0.5E-04	1.0	0.223	0.044*	0.658
88/12		-3.1	0.9E-03	8.9	-0.1E-03	6.5	0.256	0.1E-03	0.4	-0.9E-04	2.6	0.6E-04	3.0	0.309	0.005*	0.734

TABLE 3.3 CONT'D

	DF	System $VAR = \alpha_0 + \alpha_1 \ln t$					$VAR = \beta_0 + \beta_1 \ln t + \beta_2 \ln(VOL)$								
Contract	VAR $\ln(VOL)$	$\hat{\alpha}_0$	$t(\hat{\alpha}_0)$	$\hat{\alpha}_1$	$t(\hat{\alpha}_1)$	R^2	β_0	$t(\beta_0)$	β_1	$t(\beta_1)$	β_2	$t(\beta_2)$	R^2	F_1	F_2
89/1	-9.0	0.2E-03	2.6	-0.2E-04	0.7	0.005	-0.8E-03	3.4	0.8E-04	2.6	0.1E-03	4.9	0.174	0.044*	0.479
89/2	-9.6	0.3E-03	3.5	-0.3E-04	1.5	0.019	-0.1E-03	0.6	0.1E-04	0.4	0.3E-04	2.6	0.073	0.104	0.677
89/3	-10.0	0.2E-03	3.9	-0.2E-04	1.5	0.020	-0.2E-03	1.4	0.2E-04	0.9	0.4E-04	3.3	0.108	0.059	0.584
89/4	-4.6	0.5E-03	8.1	-0.1E-03	5.9	0.226	0.5E-03	3.0	-0.1E-03	4.5	-0.1E-05	0.0	0.226	0.187	0.561
89/5	-6.9	0.7E-03	8.3	-0.1E-03	6.1	0.240	0.1E-03	0.4	-0.8E-04	2.8	0.5E-04	2.7	0.286	0.071	0.719
89/6	-7.7	0.4E-03	5.9	-0.6E-04	3.5	0.091	-0.1E-03	1.0	-0.6E-05	0.2	0.5E-04	4.1	0.204	0.002*	0.691
89/7	-11.1	0.2E-01	1.7	-0.5E-02	1.5	0.019	0.2E-01	0.5	-0.5E-02	1.0	0.2E-03	0.0	0.019	0.442	0.887
89/8	-9.9	0.5E-03	5.1	-0.9E-04	3.4	0.090	0.1E-03	0.4	-0.5E-04	1.61	0.3E-04	1.6	0.109	0.187	0.508
89/9	-10.3	0.1E-03	1.2	0.1E-04	0.6	0.003	-0.4E-03	1.9	0.6E-04	2.1	0.4E-04	2.5	0.053	0.932	0.904
89/10	-9.4	0.2E-03	5.8	-0.3E-04	2.7	0.059	0.4E-03	2.9	-0.5E-04	2.7	-0.1E-04	1.1	0.069	0.474	0.190
89/11	-11.0	0.1E-03	4.3	-0.1E-04	1.2	0.013	0.3E-04	0.2	0.1E-05	0.0	0.1E-04	1.3	0.026	0.835	0.316
89/12	-11.7	0.1E-03	3.2	-0.1E-04	0.9	0.007	0.4E-05	0.0	0.2E-05	0.1	0.1E-04	1.2	0.019	0.802	0.125
90/1	-9.9	0.1E-02	4.7	-0.3E-03	4.2	0.129	0.2E-02	2.8	-0.4E-03	3.8	-0.5E-04	0.8	0.135	0.474	0.994
90/2	-4.6	0.7E-03	7.7	-0.1E-03	6.0	0.235	-0.5E-04	0.2	-0.6E-04	1.9	0.6E-04	3.7	0.317	0.145	0.852
90/3	-7.3	0.5E-03	7.0	-0.9E-04	4.9	0.175	-0.1E-03	0.7	-0.2E-04	1.0	0.5E-04	4.2	0.285	0.010*	0.336
90/4	-7.1	0.5E-03	6.1	-0.8E-04	4.0	0.116	-0.3E-03	1.7	-0.2E-04	0.9	0.8E-04	4.7	0.256	0.027*	0.719
90/5	-10.7	0.1E-02	3.8	-0.3E-03	3.2	0.078	0.2E-02	1.7	-0.3E-03	2.7	-0.5E-04	0.5	0.080	0.250	0.869
90/6	-8.1	0.3E-03	4.2	-0.4E-04	2.2	0.038	-0.3E-03	1.4	0.2E-04	0.7	0.6E-04	2.9	0.102	0.107	0.317
90/7	-7.1	0.4E-03	5.7	-0.7E-04	3.6	0.095	0.3E-04	0.1	-0.3E-04	1.1	0.3E-04	1.7	0.117	0.044*	0.284

Notes: See notes to Table 3.1.

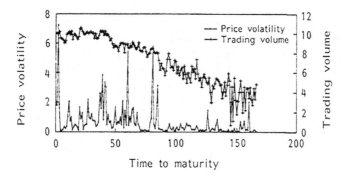

Figure 3.1: The maturity effect on crude oil 1987/11.

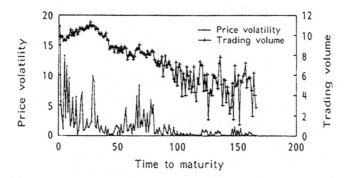

Figure 3.2: The maturity effect on crude oil 1988/11.

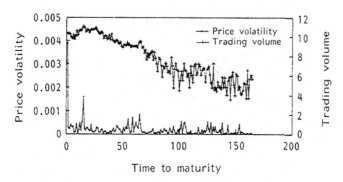

Figure 3.3: The maturity effect on crude oil 1989/5.

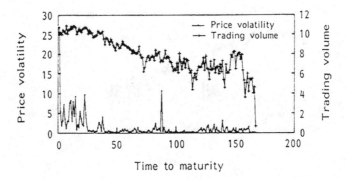

Figure 3.4: The maturity effect on crude oil 1990/2.

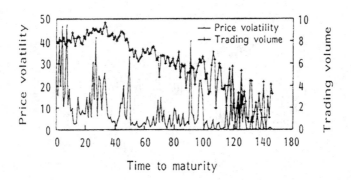

Figure 3.5: The maturity effect on heating oil 1987/.9.

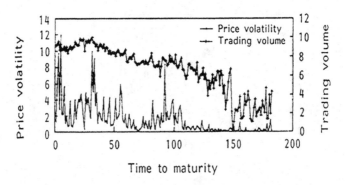

Figure 3.6: The maturity effect on heating oil 1988/12.

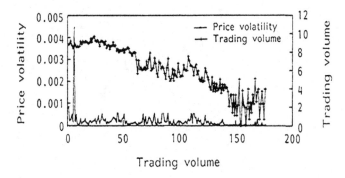

Figure 3.7: The maturity effect on heating oil 1989/4.

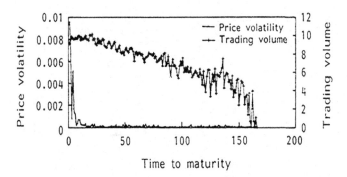

Figure 3.8: The maturity effect on heating oil 1990/1.

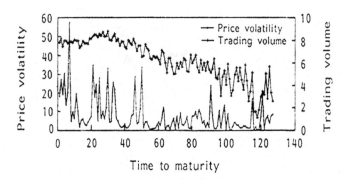

Figure 3.9: The maturity effect on unleaded gasoline 1987/9.

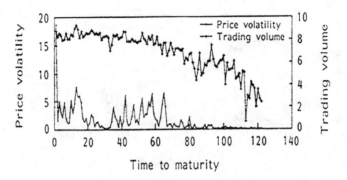

Figure 3.10: The maturity effect on unleaded gasoline 1988/10.

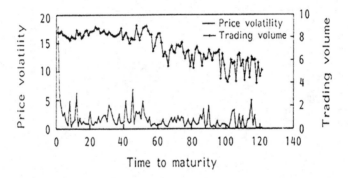

Figure 3.11: The maturity effect on unleaded gasoline 1989/4.

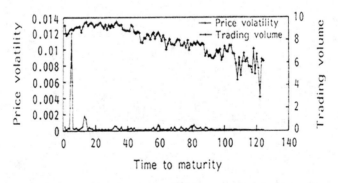

Figure 3.12: The maturity effect on unleaded gasoline 1990/5.

having degrees of freedom s and the denominator of $n - r - s - 1$. The roles of VAR and $\ln(VOL)$ are reversed in another F test to see whether there is a feedback relationship among these variables.

The results of the causality tests are displayed in the last two columns of Tables 3.1-3.3, for regressions that were run with two lag coefficients [i.e., $r = s = 2$ in equation (4)]. Tail areas (p-values) for the following asymptotic F-tests are provided in Tables 3.1-3.3. The statistic F_1 is the asymptotic F-test statistic for the null hypothesis that trading volume does not Granger cause futures price volatility. The statistic F_2 is the test statistic for the null hypothesis that futures price volatility does not Granger-cause trading volume.

Turning to the causality results, it is clear that futures price volatility does not Granger-cause trading volume (see statistic F_2), while trading volume Granger-causes futures price volatility (see statistic F_1) in 44% of the crude oil contracts, 67% of the heating oil contracts and 32% of the unleaded gasoline contracts. Clearly, these results indicate that, for a number of contracts, knowledge of past trading volume improves the prediction of futures price volatility beyond predictions that are based on past futures price volatility alone.

3.4   Conclusion

This chapter has examined the effect of maturity on the price variability of energy futures contracts. I do detect the expected negative relationship between maturity and futures price variability, but I also find that the maturity effect weakens when controlling for the effect of the volume of trade on volatility. This means that probably one or more factors simultaneously affect the volume of trade and volatility. Consequently, shedding some light on what these factors might be (that is, to investigate whether they are liquidity factors or information factors) could be the subject of particularly constructive future empirical work.

Chapter 4

Business Cycles and the Behavior of Energy Prices

*Apostolos Serletis and Vaughn W. Hulleman**

4.1   Introduction

The theory of storage, which postulates that the marginal convenience yield on inventory falls at a decreasing rate as aggregate inventory increases [see, for example, Brennan (1958), Telser (1958), and Working (1949)], is the dominant model of commodity futures prices. This hypothesis can be tested either directly by relating the convenience yield to inventory levels utilizing a simple statistical model [as in Brennan (1958) and Telser (1958)], or indirectly by testing its implication about the relative variation of spot and futures prices [as in Fama and French (1988)].

In this chapter we test the theory of storage in energy markets — crude oil, heating oil, and unleaded gas. Although the theory of storage was advanced mainly for commodities subject to seasonal variation in supply (i.e., harvest), an examination of the theory is also warranted for energy products which, although do not exhibit seasonal supply variations due to a harvest, are subject to other supply and demand seasonal fluctuations. For example, although the supply of crude oil and other refined products is not inherently seasonal, heating oil has demand peaks during the winter and gasoline has demand peaks during the summer.

---

*Originally published in *The Energy Journal* 15 (1994), 125-134. Reprinted with permission.

Early attempts at testing the theory of storage and the convenience yield hypothesis utilized industry inventory data and market prices. However, because of the difficulty in defining and measuring the relevant inventory, rather than test the theory by examining the inventory-convenience yield relation directly, we follow Fama and French (1988) and test the theory's implications about the relative variation of spot and futures prices. These implications can be viewed as refinements of Samuelson's (1965) hypothesis that [under the assumption that spot prices follow a stationary (mean-reverting) process, and futures prices are unbiased estimates for the settlement cash prices] futures prices vary less than spot prices and that the variation of futures prices is a decreasing function of maturity — see also Serletis (1992).

The remainder of the chapter consists of four sections. Section 4.1 briefly discusses the theory of storage. Section 4.2 describes the data and Section 4.3 presents the empirical results. The final section summarizes the chapter.

4.2 Theoretical Foundations

Let $F(t,T)$ be the futures price at time t for delivery of a commodity at T. Let $S(t)$ be the spot price at t and let $R(t,T)$ denote the interest rate at which market participants can borrow or lend over a period starting at date t and ending at date T. The theory of storage says [see Fama and French (1988)] that the basis — the current futures spot differential, $F(t,T) - S(t)$ — equals the interest foregone during storage, $S(t)R(t,T)$, plus the marginal warehousing cost, $W(t,T)$, minus the marginal convenience yield, $C(t,T)$. That is

$$F(t,T) - S(t) = S(t)R(t,T) + W(t,T) - C(t,T) \qquad (4.1)$$

The storage equation (4.1) is also known as the cost of carry pricing relationship and equates basis with the cost of carry, so that arbitrage is not profitable. Clearly, positive carrying costs result in a positive basis — that is, a futures price above the spot market price. In such cases the buyer of a futures contract pays a premium for deferred delivery, known as contango. Negative carrying costs imply a negative basis — that is, a futures price below the spot market price. This type of price relationship is known as backwardation. Dividing both sides of the storage equation (4.1) by $S(t)$ and rearranging, we obtain

$$\frac{F(t,T) - S(t)}{S(t)} - R(t,T) = \frac{W(t,T) - C(t,T)|}{S(t)} \qquad (4.2)$$

According to equation (4.2), the observed quantity on the left-hand side
— the interest adjusted basis, $[F(t,T) - S(t)]/S(t) - R(t,T)$ — is the
difference between the relative warehousing cost, $W(t,T)/S(t)$, and the
relative convenience yield, $C(t,T)/S(t)$.

Assuming that the marginal warehousing cost is roughly constant, that
the marginal convenience yield declines at a decreasing rate with increases
in inventory [see, for example, Brennan (1958) and Telser (1958)], and
that variation in the marginal convenience yield dominates variation in the
marginal warehousing cost, we can use the interest-adjusted-basis equation
(4.2) to develop testable hypotheses about the convenience yield. For example, when inventory is low, the relative convenience yield is high and
larger than the relative warehousing cost, and the interest-adjusted basis
becomes negative. On the other hand, when inventory is high, the relative
convenience yield falls toward zero, and the interest-adjusted basis becomes
positive and increases toward the relative warehousing cost.

Moreover, the theory of storage and the concept of declining marginal
convenience yield on inventory allow us to make predictions about the impact of demand and supply shocks on the relative variation of spot and
futures prices. For example, when inventory is high (the convenience yield
is low and the interest-adjusted basis is positive) a permanent demand
shock causes a large inventory response but a small change in the convenience yield or the interest-adjusted basis. In this case spot and futures
prices have roughly the same variability, suggesting that changes in spot
prices are largely permanent — they show up one-for-one in futures prices.
However, when inventory is low (the convenience yield is high and the
interest-adjusted basis is negative) demand shocks produce small changes
in inventories but large changes in the convenience yield and the interest-
adjusted basis. In this case, shocks cause spot prices to change more than
futures prices and the basis is more variable than when inventories are high.

In what follows, we test the theory of storage in energy markets. Because
of the difficulty, however, in defining and measuring the relevant inventory,
rather than using direct tests by relating the convenience yield to inventory
levels [see, for example, Brennan (1958) and Telser (1958)], we use the Fama
and French (1988) indirect tests based on the relative variation in spot and
futures prices. In particular, using the sign of the interest adjusted basis
as a proxy for high (+) and low (-) inventory, the prediction of the theory
of storage that shocks produce more independent variation in spot and
futures prices when inventory is low implies that the interest-adjusted basis
is more variable when it is negative — see French (1986) for a derivation
and detailed discussion.

4.3 Data

To test the theory of storage we use daily observations from the New York Mercantile Exchange (NYMEX) on one-month, two-month, four-month, and seven-month futures prices for crude oil, heating oil and unleaded gasoline. In fact, we use the spot-month futures prices as a proxy for current cash prices, the second-month futures prices as a proxy for the current futures prices, and similarly the fourth-month and seventh-month futures prices as proxies for the three-month and six-month futures prices, respectively.

The sample period is June 1, 1983 to April 27, 1992 for crude oil, August 7, 1983 to April 27, 1992 for heating oil, and December 3, 1984 to April 27, 1992 for unleaded gas. Because of daily price limits, as well as technical trading adjustments [such as, for example, abrupt movements in the price of the spot-month futures contracts on their last trading day, as traders adjust themselves out of positions], daily prices have been converted to weekly average price series. Such averaging tends to smooth these erratic price movements.

4.4 Test Results

To investigate the theory of storage prediction that the interest-adjusted basis is more variable when it is negative (because shocks produce more independent variation in spot and futures prices when inventory is low), Tables 4.1-4.3 report the standard deviations of weekly as well as daily changes in the interest-adjusted basis for one, three and six month crude oil, heating oil and unleaded gas futures contracts. Clearly, the standard deviation for crude oil is only slightly more variable when it is positive than when it is negative, but the standard deviations for heating oil and to a lesser extent for unleaded gas are larger when the interest-adjusted basis is negative. Moreover, F-tests reject (in general) the null hypothesis of equal variances. Hence, we conclude that the heating oil and unleaded gas markets pass this Fama-French (indirect) test.

As it was argued earlier, the theory of storage also implies that shocks produce roughly equal changes in spot and futures prices when inventory is high and the interest-adjusted basis is positive, but more variation in spot prices than in futures prices when inventory is low and the interest-adjusted basis is negative. To investigate this prediction of the theory of storage, we report in Table 4.4 the ratios of the standard deviation of percent futures price changes to the standard deviation of percent spot price changes for the weekly as well as the daily data. Clearly, the ratios are lower when the interest-adjusted basis is negative for all three commodities, thereby confirming the theory of storage prediction about the response of spot and futures prices to demand and supply shocks.

The theory of storage also predicts that shocks produce larger changes on shorter maturity futures prices than on longer maturity futures prices because the shocks are progressively offset by demand and supply responses. Thus, the ratios of the standard deviation of futures price changes to the standard deviation of spot price changes in Table 4.4 should decrease with increasing maturities. Clearly, the ratios are consistent with this prediction. For example, in the case of crude oil, the ratios for weekly positive interest-adjusted bases fall from 0.90 at one month to 0.85 at three months and 0.75 at six months. The ratios for negative interest-adjusted bases are 0.88, 0.78, and 0.71. This evidence is consistent with Samuelson's (1965) hypothesis about the relative variation of spot and futures prices.

TABLE 4.1
STANDARD DEVIATIONS OF CHANGES IN THE
CRUDE OIL INTEREST-ADJUSTED BASIS

Contract	Daily Data			Weekly Data		
	Positive	Negative	All	Positive	Negative	All

A. Standard Deviations of Changes in the Interest-Adjusted Basis

1-Month	0.20	0.16**	0.17	0.22	0.16**	0.17
3-Month	0.06	0.06	0.06	0.08	0.08	0.08
6-Month	0.04	0.03**	0.03	0.06	0.05^{+}	0.05

B. Number of Observations

1-Month	400	1630	2030	84	380	464
3-Month	281	1749	2030	56	407	464
6-Month	182	1848	2030	35	428	464

Notes: Statistics are for observations when the interest-adjusted basis is positive (Positive), observations when the interest-adjusted basis is negative (Negative), and for all observations (All). Interest rates in the interest-adjusted basis are yields on U.S. Treasury bills, from the Bank of Canada. Significant (rejection of the null hypothesis of equal variances) at the **one percent,*five percent, and $^{+}$ten percent level.

TABLE 4.2
STANDARD DEVIATIONS OF CHANGES
IN THE HEATING OIL INTEREST-ADJUSTED BASIS

Contract	Daily Data			Weekly Data		
	Positive	Negative	All	Positive	Negative	All

A. Standard Deviations of Changes in the Interest-Adjusted Basis

1-Month	0.14	0.21**	0.18	0.12	0.26**	0.22
3-Month	0.05	0.09**	0.08	0.06	0.14**	0.12
6-Month	0.04	0.04	0.04	0.04	0.07**	0.07

B. Number of Observations

1-Month	888	1179	2067	189	271	460
3-Month	716	1351	2067	152	308	460
6-Month	521	1546	2067	116	344	460

Notes: Statistics are for observations when the interest-adjusted basis is positive (Positive), observations when the interest-adjusted basis is negative (Negative), and for all observations (All). Interest rates in the interest-adjusted basis are yields on U.S.Treasury bills, from the Bank of Canada. Significant (rejection of the null hypothesis of equal variances) at the **one percent,*five percent, and $^+$ten percent level.

4.5 Conclusion

This chapter using the sign of the interest-adjusted basis as a proxy for high (+) and low (-) inventory, tests the prediction of the theory of storage that, when inventory is high, large inventory responses to shocks imply roughly equal changes in spot and futures prices while when inventory is low, smaller inventory responses to shocks imply larger changes in spot prices than in futures prices. Tests on spot and futures crude oil, heating oil, and unleaded gas prices confirm these predictions.

Our empirical validation of the theory of storage supports the theory's wide acceptance by market participants. In fact, as Cho and McDougall (1990, p. 611) put it

TABLE 4.3
STANDARD DEVIATIONS OF CHANGES IN THE
UNLEADED GAS INTEREST-ADJUSTED BASIS

	Daily Data			Weekly Data		
Contract	Positive	Negative	All	Positive	Negative	All

A. Standard Deviations of Changes in the Interest-Adjusted Basis

1-Month	0.21	0.20$^+$	0.20	0.21	0.21	0.21
3-Month	0.07	0.08**	0.08	0.09	0.11**	0.10
6-Month	0.03	0.04**	0.04	0.04	0.07**	0.06

B. Number of Observations

1-Month	523	1222	1745	105	267	372
3-Month	471	1274	1745	101	271	372
6-Month	311	1434	1745	65	307	372

Notes: Statistics are for observations when the interest-adjusted basis is positive
(Positive), observations when the interest-adjusted basis is negative (Negative),
and for all observations (All). Interest rates in the interest-adjusted basis are
yields on U.S. Treasury bills, from the Bank of Canada. Significant (rejection
of the null hypothesis of equal variances) at the **one percent,*five percent,
and $^+$ten percent level.

> "...the theory of storage is widely accepted by participants
> of energy futures markets. Market participants, for example,
> interpret a large negative time basis (i.e., the futures price is
> significantly lower than the spot price) as a signal to draw energy
> products out of storage and a small negative basis or positive
> basis as a signal to store commodities. Refiners frequently rely
> on basis in timing their crude oil purchases and in scheduling
> production and delivery of refined products."

Confirmation of the theory of storage is also important in modelling
futures prices. Since the theory suggests that futures prices are largely determined by demand and supply conditions in spot markets, the issue is
whether futures markets are backward or forward looking. In this regard,
Serletis and Banack (1990), using recent developments in the theory of non-
stationary regressors, test (in the context of energy markets) the hypothesis
that futures prices are unbiased predictors of future spot prices. They find

support for the hypothesis. Also, Serletis (1991) uses Fama's (1984) variance decomposition approach to measure the information in energy futures
prices about future spot prices and time varying premiums. He finds that
the premium and expected future spot price components of energy futures
are negatively correlated and that most of the variation in futures prices is
variation in expected premiums. Clearly, whether energy futures markets
are backward or forward looking is an area for potentially productive future
research.

TABLE 4.4
RATIOS OF THE STANDARD DEVIATION OF
PERCENT FUTURES PRICE CHANGES

Contract	Daily Data		Weekly Data	
	Positive	Negative	Positive	Negative
A. Crude Oil				
1-Month	0.85$^+$	0.84**	0.90	0.88
3-Month	0.79*	0.71**	0.85	0.78**
6-Month	0.69**	0.69**	0.75	0.71**
B. Heating Oil				
1-Month	0.87*	0.80**	0.93	0.83*
3-Month	0.80**	0.70**	0.83	0.75**
6-Month	0.70**	0.64**	0.73*	0.66**
C. Unleaded Gas				
1-Month	0.91	0.83**	0.93	0.90
3-Month	0.82*	0.75**	0.89	0.79*
6-Month	0.88	0.70**	0.88	0.73*

Notes: Statistics are for observations when the interest-adjusted basis is positive
(Positive), observations when the interest-adjusted basis is negative (Negative),
and for all observations (All). Interest rates in the interest-adjusted basis are
yields on U.S. Treasury bills, from the Bank of Canada. Significant (rejection of
the null hypothesis of equal variances) at the **one percent,*five percent, and
$^+$ten percent level.

Chapter 5

A Cointegration Analysis of Petroleum Futures Prices

Apostolos Serletis[*]

5.1 Introduction

One characteristic of commodity prices is the presence of a unit root in their univariate time series representation, implying that price movements are better characterized as being the sum of permanent and transitory components where the permanent component is a random walk. Although this is not a settled issue — see Perron (1989) and Serletis (1992) — and the economic significance of this distinction is a subject of continuing debate — see Cochrane (1991) and Christiano and Eichenbaum (1989) — there is also evidence that these random walk components are not different but perhaps arise from the response to the same set of fundamentals — see, for example, Baillie and Bollerslev (1989).

This chapter (partly) replicates the Baillie and Bollerslev (1989) study for spot month crude oil, heating oil and unleaded gasoline futures prices. In doing so, tests for unit roots in the univariate time series representation of daily futures prices are performed. The methodology used to study common trends in these series is based on Johansen's (1988) cointegration

---

[*]Originally published in *Energy Economics* 16 (1994), 93-97. Reprinted with permission.

framework. This is a maximum likelihood approach for estimating long-run relations in multivariate vector autoregressive models. This approach, by allowing the analysis of the data in a full system of equations model, is sufficiently flexible to account for long-run properties as well as short-run dynamics.

Cointegration is designed to deal explicitly with the analysis of the relationship between non-stationary time series. In particular, it allows individual time series to be non-stationary but requires a linear combination of the series to be stationary. Therefore, the basic idea behind cointegration is to search for linear combinations of individually non-stationary time series that are themselves stationary. Evidence to the contrary provides support for the hypothesis that the non-stationary variables have no tendency to move together over time.

The remainder of the chapter is organized as follows. Section 5.2 describes the data and analyses the univariate properties of the time series to confirm that they are not integrated of order two — a prerequisite for the analysis of cointegration. Section 5.3 outlines Johansen's (1988) multivariate approach to estimating equilibrium relationships and presents the empirical results. The chapter closes with a brief summary and conclusions.

5.2 The Data and Stochastic Trends

We study three petroleum futures markets in this chapter, those of crude oil, heating oil and unleaded gasoline. The time period of the analysis extends from 3 December 1984 to 30 April 1993, involving 2111 observations. Table 5.1 reports some summary statistics for daily returns. The skewness numbers are consistent with symmetry but the kurtosis numbers point to significant deviations from normality for all three series — there are too many large changes to be consistent with normality. The column marked $S(0)$ provides estimates of the standardized spectral density function at the zero frequency based on the Bartlett window with the window size taken to be twice the square root of the number of observations.[1] This gives

---

[1] The estimates of the standardized spectrum are comuted using the formula

$$\hat{S}(w_j) = \frac{1}{\pi} \left\{ \lambda_0 R_0 + 2 \sum_{k=1}^{m} \lambda_k R_k \cos(w_{jk}) \right\}$$

where $w_j = j\pi/m$, $j = 0, 1, \ldots, m$. m is the 'window size,' R_k is the autocorrelation coefficient of order k and λ_k is the 'lag window.' Here, I use a window size of $2\sqrt{T}$, where T is the number of observations, and Barlett's lag window, $\lambda_k = 1 - k/m$, $0 \le k \le m$. In addition, the standard errors reported for the standardized spectrum, which are valid asymptotically, are calculated as $\hat{S}(0)\sqrt{4m/3T}$.

TABLE 5.1

SUMMARY STATISTICS FOR DAILY PERCENTAGE CHANGES IN PETROLEUM FUTURES PRICES

Series	Mean	Standard Deviation	Minimum	Maximum	Skewness	Kurtosis-3	$S(0)$
Crude Oil	-.00013	.027	-.400	.140	-1.863	28.056	.920 (.221)
Heating Oil	-.00015	.027	-.390	.139	-2.540	31.782	.818 (.197)
Unleaded Gas	-.00008	.024	-.309	.123	-1.304	17.160	.724 (.174)

Notes: Sample period, daily data, 3 December 1984 to 30 April 1993 (2111 observations). $S(0)$ is a Bartlett estimate of the spectral density at zero frequency using a window size of $2\sqrt{T}$, where T is the number of observations. Numbers in parentheses are standard errors.

consistent estimates of Cochrane's (1988) measure of persistence — see, for example, Cogley (1990) — providing a useful diagnostic on the relative importance of permanent and transitory components. The point estimates suggest that all three series (and to a greater extent crude oil) contain large permanent (or random walk) components.

In Table 5.2 we also test for stochastic trends in the autoregressive representation of each individual time series. In particular, Dickey-Fuller (DF) and Augmented Dickey-Fuller (ADF) tests of the null hypothesis that a single unit root exists in the logarithm of each series are conducted using the following ADF regression:

$$\Delta \log z_t = \alpha_0 + \alpha_1 t + \alpha_2 \log z_{t-1} + \sum_{i=1}^{m} \beta_i \Delta \log z_{t-i} + e_t \qquad (5.1)$$

where z_t is the series under consideration and m is selected to be large enough to ensure that e_t is white noise. The null hypothesis of a single unit root is rejected if α_2 is negative and significantly different from zero.

TABLE 5.2

DF AND ADF TESTS FOR A UNIT ROOT IN
PETROLEUM FUTURES PRICES

Regression: $\Delta \log z_t = \alpha_0 + \alpha_1 t + \alpha_2 \log z_{t-1} + \sum_{i=1}^{m} \beta_i \Delta \log z_{t-i} + e_t$

Series	Without trend		With trend	
	DF	ADF	DF	ADF
Logarithms of the series				
Crude Oil	-2.955*	-2.605	-2.983	-2.640
Heating Oil	-3.132*	-2.704	-3.162	-2.740
Unleaded Gas	-2.796	-2.737	-2.860	-2.793
First logged differences of the series				
Crude Oil	-46.486*	-12.431*	-46.480*	-12.437*
Heating Oil	-46.482*	-12.680*	-46.475*	-12.684*
Unleaded Gas	-43.797*	-12.053*	-43.790*	-12.054*

Notes: Results are reported for a ADF statistic of order 12. The 95% critical value for the DF and ADF test statistics is -2.864 for the "without trend" version of the test and -3.414 for the "with trend" version of the test. An asterisk indicates significance at the 5% level.

In practice, the appropriate order of the autoregression in the ADF test is rarely known. One approach would be to use a model selection procedure based on some information criterion. However, Said and Dickey (1984) showed that the ADF test is valid asymptotically if the order of the autoregression is increased with sample size T at a controlled rate $T^{1/3}$. For the sample used, this translates into an order of 12. It is to be noted that for an order of zero the ADF reduces to the simple DF test. Also, the distribution of the t-test for α_2 in equation (5.1) is not standard; rather it is that given by Fuller (1976).

Table 5.2 contains DF and ADF tests of the null hypothesis that a single unit root exists in the logarithm of each series as well as in the first (logged) differences of the series. Clearly, the null hypothesis of a unit root in log levels cannot be rejected, while the null hypothesis of a second unit root is rejected. Hence, we conclude that these series are characterized as $I(1)$, i.e., having a stochastic trend. This evidence is consistent with the prevalent view that most time series are characterized by a stochastic rather than deterministic non-stationarity — see, for example, Nelson and Plosser (1982).

It is to be noted that Serletis (1992) in examining the univariate unit root properties of daily crude oil, heating oil and unleaded gasoline series (over a different sample period) using Zivot and Andrews' (1992) variation of Perron's (1989) test, shows that the unit root hypothesis can be rejected if allowance is made for the possibility of a one-time break in the intercept and the slope of the trend function at an unknown point in time. Although this has implications for both estimation and hypothesis testing, both of which rely on asymptotic distribution theory, it has no implications for the cointegration analysis that follows, since the assumption for Johansen's multivariate approach is that the series are not $I(2)$ processes — see Johansen and Juselius (1991) — which is definitely the case here.

5.3 Econometric Methodology and Empirical Results

Several methods have been proposed in the literature to estimate cointegrating vectors (long-run equilibrium relationships): see Engle and Yoo (1987) and Gonzalo (1994) for a survey and comparison. The most frequently used Engle-Granger (1987) approach is to select arbitrarily a normalization and regress one variable on the others to obtain the (OLS) regression residuals $\hat{e}$. A test of the null hypothesis of no cointegration (against the alternative of cointegration) is then based on testing for a unit root in the regression residuals $\hat{e}$ using the ADF test and the simulated critical values reported in

Engle and Yoo (1987, Table 2), which correctly take into account the number of variables in the cointegrating regression. This approach, however, does not distinguish between the existence of one or more cointegrating vectors and the OLS parameter estimates of the cointegrating vector depend on the arbitrary normalization implicit in the selection of the dependent variable in the regression equation. As a consequence, the Engle-Granger approach is well suited for the bivariate case which can have at most one cointegrating vector.

We test for the number of common stochastic trends using the multivariate approach due to Johansen (1988). This approach derives the statistical properties of the cointegration vectors by relating these vectors to the canonical correlations between the levels and first differences of the process (corrected for any short-run dynamics). Moreover, Johansen's ML approach provides relatively powerful tests — see Johansen and Juselius (1992) for a complete discussion.

In particular, following Johansen and Juselius (1992), we consider the following p $(= 3)$ dimensional vector autoregressive model:

$$\mathbf{X}_t = \sum_{i=1}^{k} \Pi_i \mathbf{X}_{t-i} + \mu + \varepsilon_t \quad (t = 1, \dots, T) \tag{5.2}$$

where $\mathbf{X}_t$ is a p-dimensional vector of petroleum futures prices and ε_t is an independently and identically distributed p-dimensional vector of innovations with zero mean and covariance matrix $\hat{Q}$. Letting $\Pi = -(\mathbf{I} - \Pi_1 - \dots - \Pi_k)$ be the $p \times p$ total impact matrix, we consider the hypothesis of the existence of at most $r(< p)$ cointegrating relations formulated as:

$$H_1(r) : \Pi = \alpha\beta' \tag{5.3}$$

where α and β are $p \times r$ matrices of full rank. The β matrix is interpreted as a matrix of cointegrating vectors, that is, the vectors in β have the property that $\beta'\mathbf{X}_t$ is stationary even though $\mathbf{X}_t$ itself is non-stationary — see Engle and Granger (1987). The α matrix is interpreted as a matrix of error correction parameters.

The maximum likelihood estimation and likelihood ratio test of this model has been investigated by Johansen (1988), and can be described as follows. First, letting $\Delta = 1 - L$, where L is the lag operator, Johansen and Juselius (1992) suggest writing equation (5.2) as

$$\Delta\mathbf{X}_t = \sum_{i=1}^{k-1} \Gamma_i \Delta\mathbf{X}_{t-i} + \alpha\beta'\mathbf{X}_{t-k} + \varepsilon_t \quad (t = 1, \dots, T) \tag{5.4}$$

where

$$\Gamma_i = -(\mathbf{I} - \Pi_1 - \dots - \Pi_i) \quad (i = 1, \dots, k-1) \tag{5.5}$$

In (5.4) the matrix $\mathbf{\Pi}$ is restricted as $\mathbf{\Pi} = \alpha\beta'$, but the parameters vary independently. Hence the parameters $\mathbf{\Gamma}_1, \ldots, \mathbf{\Gamma}_{k-1}$ can be eliminated by regressing $\Delta\mathbf{X}_t$ and $\mathbf{X}_{t-k}$ on lagged differences, $\Delta\mathbf{X}_{t-1}, \ldots, \Delta\mathbf{X}_{t-k+1}$. This gives residuals R_{ot} and R_{kt} and residual product moment matrices

$$S_{ij} = T^{-1} \sum_{t=1}^{T} \mathbf{R}_{it}\mathbf{R}'_{jt} \quad (i, j = o, k). \tag{5.6}$$

The estimate of β is found by solving the eigenvalue problem (Johansen (1988))

$$\left| \lambda S_{kk} - S_{ko}S_{oo}^{-1}S'_{ko} \right| = 0 \tag{5.7}$$

for eigenvalues $\hat{\lambda}_1 > \ldots > \hat{\lambda}_p > 0$, eigenvectors $\mathbf{V} = (\hat{v}_1, \ldots, \hat{v}_p)$ normalized by $\hat{\mathbf{V}}'S_{kk}\hat{\mathbf{V}} = \mathbf{I}$. The maximum likelihood estimators are given by

$$\hat{\beta} = (\hat{v}_1, \ldots, \hat{v}_r) \quad \hat{\alpha} = S_{ok}\hat{\beta} \quad \hat{Q} = S_{oo} - \hat{\alpha}\hat{\alpha}'. \tag{5.8}$$

Finally, the maximized likelihood function is found from

$$L_{\max}^{-2T} = \left| \hat{Q} \right| = \left| S_{oo} \right| \prod_{i=1}^{r} \left(1 - \hat{\lambda}_i \right) \tag{5.9}$$

and the likelihood ratio test of the hypothesis $H_1(r)$ is given by the trace test statistic

$$-2\log Q[H_1(r) \mid H_o] = -T \sum_{i=r+1}^{p} \log(1 - \hat{\lambda}_i). \tag{5.10}$$

An alternative test (called the maximum eigenvalue test, $\lambda_{\max}$) is based on the comparison of $H_1(r-1)$ against $H_1(r)$:

$$-2\log Q[H_1(r-1) \mid H_1(r)] = -T\log(1 - \hat{\lambda}_{r+1}). \tag{5.11}$$

Table 5.3 reports the results of the cointegration tests based on daily VARs of various lag lengths. The results for intermediate lag lengths are similar. Two test statistics are used to test for the number of cointegrating vectors: the trace and maximum eigenvalue ($\lambda_{\max}$) test statistics. In the trace test the null hypothesis that there are at most r cointegrating vectors where $r = 0$, 1 and 2 is tested against a general alternative whereas in the maximum eigenvalue test the alternative is explicit. That is, the null hypothesis $r = 0$ is tested against the alternative $r = 1$, $r = 1$ against the alternative $r = 2$, etc. The 95% critical values of the trace and maximum eigenvalue test statistics are taken from Johansen and Juselius (1990, Table A2).

TABLE 5.3
JOHANSEN TESTS FOR COINTEGRATION AMONG PETROLEUM FUTURES PRICES

H_0	$k=2$ Trace	λ_{max}	$k=4$ Trace	λ_{max}	$k=6$ Trace	λ_{max}	Critical values Trace 95%	90%	λ_{max} 95%	90%
$r=0$	86.936*	56.820*	84.305*	56.153*	73.564*	45.369*	31.256	28.436	21.279	18.959
$r \leq 1$	30.116*	22.298*	28.152*	21.767*	28.195*	21.935*	17.844	15.583	14.595	12.783
$r \leq 2$	7.818	7.818	6.384	6.384	6.259	6.259	8.083	6.691	8.083	6.691

Notes: Critical values are from Johansen and Juselius (1988, Table A2); k refers to the number of lags in the VAR. Drift maintained. Eigenvalues for $k = 3$, (0.026 0.010 0.003). Eigenvalues for $k = 4$, (0.026 0.010 0.003). Eigenvalues for $k = 6$, (0.021 0.010 0.002). An asterisk indicates significance at the 5% level.

Clearly, the two test statistics give similar results. In particular, the hypothesis of one or less cointegrating relations has to be rejected. Thus, our three variables form two cointegrating relationships, or alternatively they are driven by only one common trend. Under the common trends interpretation — see, for example, Stock and Watson (1988) — this result is not too surprising. The same underlying stochastic components presumably affect all petroleum futures markets.

5.4   Conclusion

The maximum likelihood cointegration analysis of daily spot-month crude oil, heating oil and unleaded gasoline futures prices covering the period 3 December 1984 to 30 April 1993 led to the conclusion that all three spot-month futures prices are driven by only one common trend, suggesting that it is appropriate to model energy futures prices as a cointegrated system. Further research may suggest a useful way of identifying the common non-stationary factor so that it can be estimated and studied. This will undoubtedly improve our understanding of how petroleum futures prices change over time.

Part 2

Natural Gas Markets

Overview of Part 2

Apostolos Serletis

The following table contains a brief summary of the contents of the chapters in Part 2 of the book. This part of the book consists of three chapters addressing a number of issues regarding natural gas markets.

Natural Gas Markets

Chapter Number	Chapter Title	Contents
6	Is There an East-West Split in North American Natural Gas Markets?	It uses the Engle and Granger (1987) approach for estimating bivariate cointegrating relationships as well as Johansen's (1988) maximum likelihood approach to present evidence that an east-west split in North American natural gas markets does not exist.
7	Business Cycles and Natural Gas Prices	This chapter investigates the basic stylized facts on natural gas price movements, using the methodology suggested by Kydland and Prescott (1990) as well as Granger (1969) causality tests. It shows that natural gas prices are procyclical and lag the cycle of industrial production in the United States.
8	Futures Trading and the Storage of North American Natural Gas	Chapter 8 tests the theory of storage in North American natural gas markets using the Fama and French (1988) indirect test. It confirms the predictions of the theory of storage.

Chapter 6:

This chapter presents evidence concerning shared stochastic trends in North American natural gas (spot) markets, using monthly data for the period that natural gas has been traded on organized exchanges (from June, 1990 to January, 1996). In doing so, it uses the Engle and Granger (1987) approach for estimating bivariate cointegrating relationships as well as Johansen's (1988) maximum likelihood approach for estimating cointegrating relationships in multivariate vector autoregressive models. The results indicate that the east-west split does not exist.

Chapter 7:

This chapter investigates the basic stylized facts of natural gas price movements using the methodology suggested by Kydland and Prescott (1990). The results indicate that natural gas prices are procyclical and lag the cycle of industrial production. Moreover, natural gas prices are positively contemporaneously correlated with U.S. consumer prices and lead the cycle of consumer prices, raising the possibility that natural gas prices might be a useful guide for U.S. monetary policy, like crude oil prices are, possibly serving as an important indicator variable.

Chapter 8:

This Chapter tests the theory of storage in North American natural gas markets, using the Fama and French (1988) indirect test. In particular, it tests the prediction of the theory that, when inventory is high, large inventory responses to shocks imply roughly equal changes in spot and futures prices, whereas when inventory is low, smaller inventory responses to shocks imply larger changes in spot prices than in futures prices. The tests on spot and futures North American natural gas prices confirm these predictions of the theory of storage.

Chapter 6

Is There an East-West Split in North American Natural Gas Markets?

Apostolos Serletis[*]

6.1 Introduction

In the last decade, the North American natural gas industry has seen a dramatic transformation from a highly regulated industry to one which is more market-driven. The transition to a less regulated, more market-oriented environment has led to the emergence of different spot markets throughout North America. In particular, producing area spot markets have emerged in Alberta, British Columbia, Rocky Mountain, Anadarko, San Juan, Permian, South Texas, and Louisiana basins. Moreover, production sites, pipelines and storage services are more accessible today, thereby ensuring that changes in market demand and supply are reflected in prices on spot, futures, and swaps markets.

In a perfectly competitive industry the law of one price suggests that the difference in prices between any two markets should reflect the difference in transportation costs between the two markets. Because the natural gas molecule is identical when measured in terms of heating values, whether it comes from a well in Alberta or in the Gulf Coast, there is no reason that

---

*Originally published in *The Energy Journal* 18 (1997), 47-62. Reprinted with permission.

the law of one price should not apply to the natural gas industry. However, capacity constraints seem to be distorting North American natural gas markets in such a way that varying differentials emerge between spot prices, reflecting not only transportation costs but also supply and demand conditions in different areas.

Recently, King and Cuc (1996), in investigating the degree of natural gas spot price integration in North America, report evidence of an east-west split in North American natural gas markets. In particular, they argue that western prices tend to move together and similarly eastern prices tend to move together, but there seems to be a divergence between eastern and western prices. In other words, according to King and Cuc (1996) eastern and western prices seem to be determined by different fundamentals. King and Cuc (1996) use integration and (bivariate) cointegration analysis to measure natural gas price convergence, but mainly rely on a method of measuring convergence recently proposed by Haldane and Hall (1991). This method is based on the use of time-varying parameter (Kalman filter) analysis and is typically used to estimate regression type-models where the coefficients follow a random process over time.

In this chapter we investigate the robustness of the King and Cuc (1996) findings to alternative testing methodologies. In doing so, we test for shared stochastic price trends using current, state-of-the-art econometric methodology. In particular, we pay explicit attention to the time series properties of the variables and test for cointegration, using both the Engle and Granger (1987) approach as well as Johansen's (1988) (multivariate) maximum likelihood extension of the Engle and Granger approach. Looking ahead to the results, the tests indicate that the King and Cuc (1996) east-west split does not exist.

The chapter is organized as follows. Section 6.2 provides some background regarding North American natural gas spot markets. Section 6.3 discusses the data and investigates the univariate time series properties of the variables, since meaningful cointegration tests critically depend on such properties. Section 6.4 tests for cointegration and presents the results. The last section concludes the chapter.

6.2   The North American Natural Gas Spot Markets

The Alberta and British Columbia producing regions are part of the Western Canadian Sedimentary basin. In the case of Alberta, natural gas is transported from the field along the Nova Gas Transmission system for sale within Alberta as well as exported to eastern Canada and the United States.

The British Columbia natural gas producing region is located mainly in northeastern British Columbia and natural gas is transported from the field along the Westcoast Gas Services system for sale in British Columbia and for export to the United States. Whereas gas exported from Alberta is resold in eastern markets in Canada and in the northeastern and midwestern United States, as well as in the western United States (specifically in California and the Pacific Northwest), British Columbia exports generally serve only markets in the Pacific Northwest and California.

The Rocky Mountain basin is a cluster of producing regions in the states of Wyoming, Utah, and Colorado. Pipelines in this area can transport production either east or west, although the eastward capacity has been constrained and thus, the majority of Rocky Mountain supplies are sold in the western markets in California and the Pacific Northwest. The San Juan basin is located in southwestern Colorado and northwestern New Mexico. Like the Rocky Mountain basin, eastward capacity is constrained which means that the majority of gas produced in this region is sold in western markets. The Rocky Mountain, San Juan, and Western Canadian Sedimentary basins comprise the western portion of the King and Cuc's (1996) east-west split (see Figure 6.1).

In the next two sections, we investigate whether the price behavior of natural gas in different areas is similar. In particular, we use recent advances in the theory of nonstationary regressors to determine what trends in natural gas prices, if any, are common to Alberta, British Columbia, Rocky Mountain, San Juan, Anardarko, Louisiana, Permian and South Texas? Our definition of trend follows the cointegration literature. In particular, according to Beveridge and Nelson (1981) any time series characterized by a unit root can be decomposed into a random-walk and a stationary component, with the random-walk component being interpreted as the stochastic trend. Two series are said to share a trend if their stochastic trend components are proportional to each other. Clearly, a better understanding of the extent to which natural gas prices share trends might shed some light on the economic processes that determine natural gas prices.

6.3 The Data and Stochastic Trends

The data we use to test for shared stochastic natural gas price trends (from June 1990 to January 1996) are monthly bid-week prices reported by Brent Friedenberg Associates in the Canadian Natural Gas Focus. Bid week refers to the week during which pipeline nominations to transport gas take place. This is generally five days before the end of the month. Figure 6.2 shows the

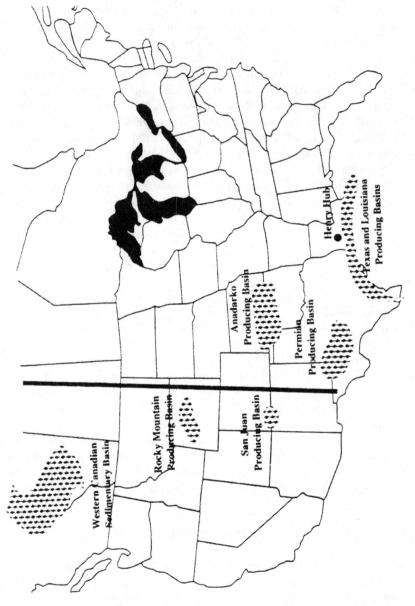

Figure 6.1: The North American Natural Gas Industry — An East-West
Split?

plots of natural gas prices in the western producing region of the King and Cuc (1996) east-west split — Alberta, British Columbia, Rocky Mountain, and San Juan basins. Figure 6.3 shows prices in the eastern producing region of the split — Anadarko, Louisiana, Permian, and South Texas basins.

The first step in testing for shared stochastic trends is to test for stochastic trends (unit roots) in the autoregressive representation of each individual time series. Nelson and Plosser (1982) argue that most macroeconomic and financial time series have a unit root (a stochastic trend), and describe this property as one of being "difference stationary" so that the first difference of a time series is stationary. An alternative "trend stationary" model has been found to be less appropriate. In what follows we test for unit roots using three alternative unit root testing procedures to deal with anomalies that arise when the data are not very informative about whether or not there is a unit root. In doing so, we choose to include only a constant (but not a time trend), since the series are not trending (see Figures 6.2 and 6.3).

The first three columns of panel A of Table 6.1 report p-values for the augmented Weighted Symmetric (WS) unit root test [see Pantula *et al.* (1994)], the augmented Dickey-Fuller (ADF) test [see Dickey and Fuller (1981)], and the $Z(t_{\hat{a}})$ nonparametric test of Phillips (1987) and Phillips and Perron (1988). These p-values are based on the response surface estimates given by MacKinnon (1994). For the WS and ADF tests, the optimal lag length was taken to be the order selected by the Akaike information criterion (AIC) plus 2 — see Pantula *et al.* (1994) for details regarding the advantages of this rule for choosing the number of augmenting lags. The $Z(t_{\hat{a}})$ test is done with the same Dickey-Fuller regression variables, using no augmenting lags. Based on the p-values for the WS, ADF, and $Z(t_{\hat{a}})$ test statistics reported in panel A of Table 6.1, the null hypothesis of a unit root in log levels cannot be rejected, except perhaps for the Permian price series.

To test the null hypothesis of a second unit root, in panel B of Table 6.1 we test the null hypothesis of a unit root [using the WS, ADF, and $Z(t_{\hat{a}})$ tests] in the first (logged) differences of the series. Clearly, all the series appear to be stationary in growth rates, since the null hypothesis of a unit root in the first (logged) differences of the series is rejected. We conclude that all the series are integrated of order one [or $I(1)$ in the terminology of Engle and Granger (1987)].

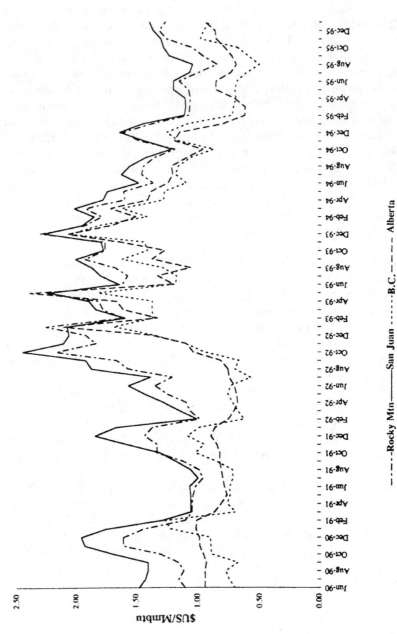

Figure 6.2: Western Prices

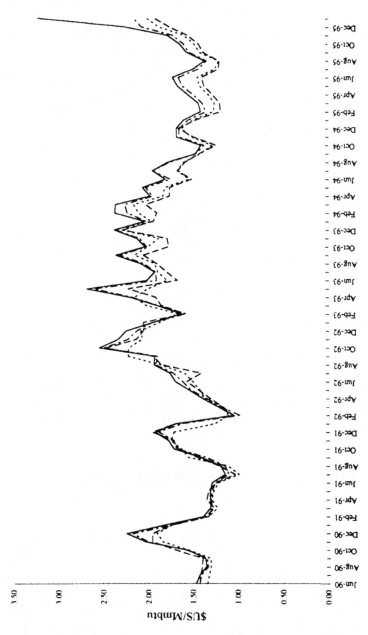

Figure 6.3: Eastern Prices

TABLE 6.1
MARGINAL SIGNIFICANCE LEVELS OF UNIT ROOT TESTS
IN NORTH AMERICAN NATURAL GAS SPOT PRICES

Market	A. Log Levels			B. First differences of log levels		
	WS	ADF	$Z(t_{\hat{\alpha}})$	WS	ADF	$Z(t_{\hat{\alpha}})$
Rocky Mountain	.1107	.218	.110	.000	.000	.000
San Juan	.030	.127	.065	.000	.002	.000
Permian	.010	.044	.020	.000	.001	.000
Anadarko	.063	.139	.074	.000	.000	.000
South Texas	.078	.055	.031	.000	.000	.000
Louisiana	.127	.219	.060	.000	.000	.000
British Columbia	.113	.232	.099	.000	.000	.000
Alberta	.092	.274	.158	.000	.000	.000

Notes: Tests use a constant (but not a time trend). Numbers are tail areas of unit root tests. The number of augmenting lags is determined using the AIC+2 rule. p-values less than 0.05 reject the null hypothesis of a unit root at the 0.05 level of significance.

6.4 Test Methods (and Capabilities) and Results

Since a stochastic trend has been confirmed for each price series, we now explore for shared stochastic price trends among these series by testing for cointegration (i.e., long-run equilibrium relationships). Cointegration is a relatively new statistical concept designed to deal explicitly with the analysis of the relationship between nonstationary time series. In particular, it allows individual time series to be nonstationary, but requires a linear combination of the series to be stationary. Therefore, the basic idea behind cointegration is to search for a linear combination of individually nonstationary time series that is itself stationary. Evidence to the contrary provides strong empirical support for the hypothesis that the integrated variables have no inherent tendency to move together over time.

Several methods have been proposed in the literature to estimate cointegrating vectors — see Engle and Yoo (1987) and Gonzalo (1994) for a survey and comparison. The most frequently used Engle-Granger (1987) approach is to select arbitrarily a normalization and regress one variable on the others to obtain the ordinary least squares (OLS) regression residuals $\hat{e}$. A test of the null hypothesis of no cointegration (against the alternative of cointegration) is then based on testing for a unit root in the regression residuals $\hat{e}$ using the ADF test and simulated critical values which correctly take into account the number of variables in the cointegrating regression. This approach, however, does not distinguish between the existence of one or more cointegrating vectors and the OLS parameter estimates of the cointegrating vector depend on the arbitrary normalization implicit in the selection of the dependent variable in the regression equation. As a consequence, the Engle-Granger approach is well suited for the bivariate case which can have at most one cointegrating vector.

Table 6.2 reports asymptotic p-values [computed using the coefficient estimates in MacKinnon (1994)] of bivariate cointegration tests (in log levels). The entries across each row are the p-values for testing the null of no cointegration between the variable indicated in the row heading and the variable indicated in the column heading, with the variable indicated in the row heading being the dependent variable. In other words, the cointegration tests are first done with one series as the dependent variable in the cointegrating regression and then with the other series as the dependent variable — we should be wary of a result indicating cointegration using one series as the dependent variable, but no cointegration when the other series is used as the dependent variable. This possible ambiguity is a weakness of the Engle Granger cointegration test. The tests are a constant (but not a trend variable) and the number of augmenting lags is chosen using the AIC+2 rule mentioned earlier.

The results suggest that the null hypothesis of no cointegration cannot be rejected (at the 5 percent level), except for the pairs Rocky Mountain-San Juan, Permian-Anadarko, Anadarko-Alberta, Alberta-South Texas, and Alberta-British Columbia. That is, only five out of the twenty-eight price pairs cointegrate (at the 5 percent level and none at the 1 percent level). These results are very different from those reported in King and Cuc (1996) — there is much less cointegration across series indicating anything but an east-west split. The difference is due to the different time period than that considered in King and Cuc (1996) and possibly due to the inclusion of a trend variable in their cointegrating regressions, which reduced degrees of freedom and the power of the test — reduced power means that they conclude that the series cointegrate when in fact they don't. Notice that the King and Cuc (1996) study is not clear on how deterministic components in the time series were treated.

To investigate the robustness of these results to alternative testing methodologies, (under the assumption that North American natural gas prices are determined simultaneously) we consider the joint modelling of these prices and test for shared stochastic trends using Johansen's (1988) maximum likelihood extension of the Engle and Granger approach. Johansen's maximum likelihood approach to the estimation of the number of linearly independent cointegrating vectors for a vector autoregressive process, X_t, of order p involved (i) regression ΔX_i on $\Delta X_{t-1}, \ldots, \Delta X_{t-p+1}$, (ii) regressing ΔX_{t-p} on the same set of regressors, and (iii) performing a canonical correlation analysis on the residuals of these two regressions — see Johansen (1988) for more details or Serletis (1994) for an application.

We search for shared stochastic price trends among prices within two price groups — eastern and western. If any shared trends are found in the eastern (western) price group [as King and Cuc (1996) suggest], then they might sensibly be thought of as the eastern (western) natural gas price trends. In fact, according to King and Cuc (1996), prices within each price group tend to move together, responding to the same set of fundamentals, meaning that there is one shared stochastic price trend within each price group. Using the Engle and Granger (1987) terminology, we say that in an n-variable system with m cointegrating vectors there are $n - m$ common trends.

Tables 6.3 and 6.4 report the results of the cointegration tests based on monthly VARs of various lag lengths for the eastern and western price groups, respectively. The results for intermediate lag length are similar. Two test statistics are used to test for the number of cointegrating vectors: the trace and maximum eigenvalue ($\lambda_{\max}$) test statistics. In the trace test the null hypothesis that there are at most r cointegrating vectors where $r = 0, 1, 2$, and 3 is tested against a general alternative whereas in the maximum eigenvalue test the alternative is explicit. That is, the null hypothesis $r = 0$ is tested against the alternative $r = 1$, $r = 1$ against the alternative $r = 2$, etc. The 95 percent critical values of the trace and maximum eigenvalue test statistics are taken from Osterwald-Lenum (1992).

Clearly, the two test statistics give similar results in both Tables 6.3 and 6.4. In particular, the trace and $\lambda_{\max}$ statistics reject [at conventional significance levels, based on the critical values reported by Osterwald-Lenum (1992)] the null hypothesis of no cointegrating vectors ($r = 0$) and accept the alternative of one or more cointegrating vectors. However, the null of $r \leq 1$ cannot be rejected, indicating no more than one cointegrating vector within each natural gas price group. Hence, confirming the impression

TABLE 6.2

MARGINAL SIGNIFICANCE LEVELS OF BIVARIATE ENGLE-GRANGER (1987)
COINTEGRATION TESTS BETWEEN NORTH AMERICAN NATURAL GAS SPOT PRICES

	Rocky Mountain	San Juan	Permian	Anadarko	South Texas	Louisiana	British Columbia	Alberta
Rocky Mountain	—	.032	.641	.434	.720	.978	.215	.008
San Juan	.017	—	.930	.122	.752	.968	.112	.156
Permian	.185	.627	—	.003	.364	.684	.151	.048
Anadarko	.142	.048	.018	—	.637	.958	.097	.043
South Texas	.590	.476	.517	.726	—	.936	.067	.023
Louisiana	.973	.972	.894	.983	.971	—	.283	.170
British Columbia	.225	.211	.292	.246	.286	.572	—	.002
Alberta	.226	.259	.443	.048	.045	.234	.008	—

Notes: All tests use a constant (but not a trend variable). The number of lags is determined using the AIC+2 rule. Asymptotic p-values are computed using the coefficients in MacKinnon (1994). Low p-values imply strong evidence against the null of no cointegration.

TABLE 6.3
Johansen Tests for Cointegration among Eastern
(Anadarko, Louisiana, Permian, & South Texas) Natural Gas Prices

H_0	$k = 2$		$k = 4$		$k = 6$		Critical Values			
							Trace		λ_{max}	
	Trace	λ_{max}	Trace	λ_{max}	Trace	λ_{max}	95%	90%	95%	90%
$r = 0$	68.068*	45.368*	46.285	32.386*	49.066	31.216*	53.116	49.648	28.138	25.559
$r \leq 1$	22.699	11.007	13.898	7.528	17.850	9.235	34.910	32.003	22.002	19.766
$r \leq 2$	11.692	9.727	6.370	4.651	8.615	5.774	19.964	17.852	15.672	13.752
$r \leq 3$	1.964	1.964	1.718	1.718	2.840	2.840	9.243	7.525	9.243	7.525

Notes: Critical values are from Osterwald-Lenum (1992). k refers to the number of lags in the VAR. Drift maintained. An asterisk indicates significance at the 5% level.

TABLE 6.4

JOHANSEN TESTS FOR COINTEGRATION AMONG WESTERN
(ALBERTA, BRITISH COLUMBIA, ROCKY MOUNTAIN, AND SAN JUAN) NATURAL GAS PRICES

H_0	$k=2$		$k=4$		$k=6$		Critical Values			
	Trace	λ_{max}	Trace	λ_{max}	Trace	λ_{max}	Trace		λ_{max}	
							95%	90%	95%	90%
$r=0$	63.323*	29.779*	64.252*	30.921*	68.283*	29.667*	53.116	49.648	28.138	25.559
$r\leq1$	33.526	15.315	33.331	15.252	13.616*	22.814	34.910	32.003	22.002	19.766
$r\leq2$	18.211	13.252	17.805	13.185	15.801	12.632	19.964	17.852	15.672	13.752
$r\leq3$	4.959	4.959	4.620	4.620	3.169	3.169	9.243	7.525	9.243	7.525

Notes: Critical values are from Osterwald-Lenum (1992). k refers to the number of lags in the VAR. Drift maintained. An asterisk indicates significance at the 5% level.

from Table 6.2, natural gas spot prices within each price group respond to different underlying stochastic components.

When interpreting the results in terms of convergence, it should be noted that cointegration analysis cannot in principle detect convergence, because it fails to take account of the fact that convergence is a gradual and on-going process, which implies that statistical tests should lead to reject the null hypothesis of no cointegration only when convergence has already taken place — see, for example, Bernard (1992). In other words, the tests conducted here are tests for convergence over the whole period under consideration, but these tests are not tests of a move from non-convergence to convergence — the latter being the issue that King and Cuc (1996) mainly investigate.

6.5   Conclusion

This chapter explored the behavior of North American natural gas price trends and their interrelations. The degree of shared trends among natural gas prices is of considerable importance. For example, if natural gas prices share trends, in the sense that their stochastic trend components are proportional to each other, then natural gas markets have an error-correction mechanism — that is, every permanent shock in one market is ultimately transmitted to the other markets.

We applied the Engle and Granger (1987) two-step procedure to bivariate natural gas price relationships and we also tested for the number of common stochastic trends among prices within eastern and western markets using the powerful multivariate approach due to Johansen (1988). The results led to the conclusion that natural gas prices do not cointegrate and that, in particular, natural gas prices within each area (eastern and western) are driven by different stochastic trends, meaning that the east-west split does not exist.

One way to interpret these results is in terms of the absence or presence of unexploited profit opportunities. In the case of integrated price series that cointegrate, the price differential is stationary, implying price convergence, a high degree of price competition, and the absence of unexploited profit opportunities. In this case, every permanent shock in the trend of one series is ultimately transmitted to the trend of the other series. In the case, however, of integrated price series that do not cointegrate (which is the case of North American natural gas spot prices), the difference between the respective prices fluctuates stochastically, in excess of transmission and transaction costs, indicating the failure of potential arbitrage to discipline prices. In this case, the marginal value of the commodity across locations differs by more than transmission and transaction costs suggesting unexploited profit opportunities.

Chapter 7

Business Cycles and Natural Gas Prices

*Apostolos Serletis and Asghar Shahmoradi**

7.1 Introduction

In recent years, the North American energy industry has undergone major structural changes that have significantly affected the environment in which producers, transmission companies, utilities and industrial customers operate and make decisions. For example, major policy changes are the U.S. Natural Gas Policy Act of 1978, Natural Gas Decontrol Act of 1989, and FERC Orders 486 and 636. In Canada, deregulation in the mid-1980s has also broken the explicit link between the delivered prices of natural gas and crude oil (that was in place prior to 1985), and has fundamentally changed the environment in which the Canadian oil and gas industry operates. Moreover, the Free Trade Agreement (FTA) signed in 1988 by the United States and Canada, and its successor, the North American Free Trade Agreement (NAFTA) signed in 1993 by the United States, Canada, and Mexico, have underpinned the process of deregulation and attempted to increase the efficiency of the North American energy industry.

In this chapter we systematically investigate the cyclical behavior of natural gas price movements for the period that natural gas has been traded on an organized exchange. The cyclical behavior of energy prices, in general, is important and has been the subject of a large number of studies, exemplified by Hamilton (1983). These studies have, almost without exception,

---

*Originally published in *OPEC Review* (2005), 75-84. Reprinted with permission.

concentrated on the apparently adverse business-cycle effects of oil price shocks. For example, Hamilton (1983) working on pre-1972 data and based on vector autoregression (VAR) analysis, concluded that energy prices are countercyclical and lead the cycle. More recently, however, Serletis and Kemp (1998) show, using data over the period for which energy has been traded on organized exchanges and the methodology suggested by Kydland and Prescott (1990), that energy prices are in general procyclical.

The chapter is organized as follows. Section 7.2 uses the Hodrick and Prescott (1980) and Baxter and King (1999) filtering procedures for decomposing time series into long-run and business cycle components and presents empirical correlations of natural gas prices with U.S. industrial production and consumer prices, as well as with West Texas Intermediate (WTI) crude oil, heating oil, and propane prices. Section 7.3 tests for Granger causality, explicitly taking into account the univariate and bivariate properties of the variables. The final section summarizes the chapter.

7.2   The Stylized Facts

In this section we investigate the basic stylized facts of natural gas price movements, using stationary cyclical deviations based on the Hodrick and Prescott (1980) and the Baxter and King (1999) filters; see Hodrick and Prescott (1980) and Baxter and King (1999) for more details regarding these filters. In doing so, we use monthly data from January 1990 to March 2002 (a total of 147 monthly observations) and define natural gas cycle regularities as the dynamic comovements of the cyclical component of natural gas prices and the cycle. In particular, the business cycle regularities that we consider are autocorrelations and dynamic cross-correlations between the cyclical component of natural gas prices, on the one hand, and the cyclical component of U.S. industrial production on the other.

We measure the degree of comovement of natural gas prices with the cycle by the magnitude of the correlation coefficient $\rho(j)$, $j \in \{0, \pm1, \pm2, \ldots\}$. The contemporaneous correlation coefficient — $\rho(0)$ — gives information on the degree of contemporaneous comovement. In particular, if $\rho(0)$ is positive, zero, or negative, we say that the series is procyclical, acyclical, or countercyclical, respectively. In fact, following Fiorito and Kollintzas (1994), for $0.23 \leq |\rho(0)| < 1$, $0.10 \leq |\rho(0)| < 0.23$, and $0 \leq |\rho(0)| < 0.10$, we say that the series is strongly contemporaneously correlated, weakly contemporaneously correlated, and contemporaneously uncorrelated with the cycle, respectively.[1] The cross correlation coefficient, $\rho(j)$, $j \in \{\pm1, \pm2, \ldots\}$,

---

[1] The cutoff point of 0.1 is close to the value of 0.097 that is required to reject the null hypothesis $H_0 : \rho(0) = 0$ at the 5% level. Also, the cutoff point of 0.23

gives information on the phase shift of natural gas relative to the cycle. If $|\rho(j)|$ is maximum for a positive, zero, or negative j, we say that the cycle of natural gas prices is leading the cycle by j periods, is synchronous, or is lagging the cycle by j periods, respectively.

Table 7.1 reports the contemporaneous correlations as well as the cross correlations based on the Hodrick-Prescott (Panel A) and Baxter-King (Panel B) filters, at lags and leads of 1, 2, 3, 6, 9, and 12 months, between the cyclical component of spot Henry Hub natural gas prices and the cyclical component of each of U.S. industrial production, U.S. consumer prices, West Texas Intermediate crude oil prices, heating oil prices, and propane prices. The industrial production and consumer price indexes were obtained from the Federal Reserve Economic Database (FRED), maintained by the Federal Reserve Bank of St. Louis (http://research.stlouisfed.org/fred/index.html). The spot crude oil and natural gas prices were obtained from the Oil & Gas Journal's database (http://orc.pennnet.com/home.cfm). Finally, the spot heating oil and propane prices were obtained from the U.S. Energy Information Administration (http://www.eia.doe.gov).

Clearly, irrespective of the filter used, natural gas prices are procyclical and lag the cycle (of industrial production). This is consistent with the evidence reported by Serletis and Kemp (1998) using spot-month NYMEX natural gas futures prices (as a proxy for the spot price) over a much shorter sample period (with only 37 monthly observations). Moreover, (regardless of which filter is used) natural gas prices are positively contemporaneously correlated with U.S. consumer prices and the cycle of natural gas leads the cycle of consumer prices, suggesting that changes in natural gas prices might be good predictors of future aggregate price changes. Finally, the contemporaneous correlation of natural gas prices is strikingly strong with crude oil, heating oil, and to a larger extent with propane, suggesting that these markets are perhaps driven by one common trend — see Serletis (1994) for work along these lines.

In the next section we investigate whether the apparent phase-shift between natural gas prices and each of the other variables justifies a causal relationship between these variables. In doing so, we interpret causality in terms of predictability and not as suggesting the existence of underlying structural relationships between the variables.

---

is close to the value of 0.229 that is required to reject the null $H_0 : |\rho(0)| \leq 0.5$ at the 5% level.

TABLE 7.1
CYCLICAL CORRELATIONS OF NATURAL GAS PRICES WITH INDUSTRIAL PRODUCTION, CONSUMER PRICES, CRUDE OIL, HEATING OIL, AND PROPANE

$\rho(x_t, y_{t+j})$, $j = -12, -9, -6, -3, -2, -1, 0, 1, 2, 3, 6, 9, 12$

	$j=-12$	$j=-9$	$j=-6$	$j=-3$	$j=-2$	$j=-1$	$j=0$	$j=1$	$j=2$	$j=3$	$j=6$	$j=9$	$j=12$
				Panel A. Hodrick and Prescott Filter									
Industrial production	-.033	.150	.354	.419	.403	.376	.338	.271	.212	.131	-.128	-.324	-.329
Consumer prices	-.182	.043	.138	.282	.369	.473	.578	.667	.670	.640	.429	.181	-.099
Crude oil	.160	.353	.365	.457	.488	.508	.514	.487	.428	.350	.106	-.110	-.283
Heating oil	.118	.307	.344	.539	.585	.610	.612	.553	.476	.386	.115	-.110	-.308
Propane	-.012	.134	.242	.467	.562	.675	.734	.631	.485	.390	.171	.018	-.283
				Panel B. Baxter and King Band-Pass Filter									
Industrial production	.102	.328	.478	.478	.444	.393	.322	.240	.159	.083	-.081	-.137	-.142
Consumer prices	-.035	.120	.205	.332	.406	.492	.582	.599	.589	.558	.369	.170	.017
Crude oil	.430	.458	.381	.349	.360	.375	.384	.365	.332	.287	.110	-.057	-.182
Heating oil	.337	.455	.482	.507	.517	.525	.523	.483	.427	.359	.120	-.090	-.234
Propane	.251	.265	.289	.442	.510	.572	.615	.590	.537	.467	.220	.027	-.113

Note: Results are reported using monthly data for the period January 1990 to March 2002. x_t =Natural gas, y_t =(Industrial production,Consumer prices, Crude oil, Heating oil, Propane).

7.3   Granger Causality Tests

The first step in testing for causality is to test for the presence of a stochastic trend in the autoregressive representation of each (logged) individual time series. In the first three columns of Table 7.2 we report p-values [based on the response surface estimates given by MacKinnon (1994)] for the weighted symmetric (WS) unit root test [see Pantula, Gonzalez-Farias, and Fuller (1994)], the augmented Dickey-Fuller (ADF) test [see Dickey and Fuller (1981) for more details], and the nonparametric $Z(t_{\hat{\alpha}})$ test of Phillips and Perron (1987). As discussed in Pantula et $al.$ (1994), the WS test dominates the ADF test in terms of power. Also the $Z(t_{\hat{\alpha}})$ test is robust to a wide variety of serial correlation and time-dependent heteroskedasticity. For the WS and ADF tests, the optimal lag length is taken to be the order selected by the Akaike Information Criterion (AIC) plus 2 — see Pantula et $al.$ (1994) for details regarding the advantages of this rule for choosing the number of augmenting lags. The $Z(t_{\hat{\alpha}})$ test is done with the same Dickey-Fuller regression variables, using no augmenting lags. According to these p-values, the null hypothesis of a unit root in log levels cannot be rejected except for heating oil, suggesting that these series are integrated of order 1 [or I(1) in the terminology of Engle and Granger (1987)].[2]

Next we explore for shared stochastic trends between natural gas prices and each of the other I(1) variables using methods recommended by Engle and Granger (1987). That is, we test for cointegration (i.e., long-run equilibrium relationships). If the variables are I(1) and cointegrate, then there is a long-run equilibrium relationship between them. Moreover, the dynamics of the cointegrated variables can be described by an error correction model, in which the short-run dynamics are influenced by the deviation from the long-run equilibrium. If, however, the variables are I(1) but do not cointegrate, ordinary least squares yields misleading results. In that case, the only valid relationship that can exist between the variables is in terms of their first differences.

We test the null hypothesis of no cointegration (against the alternative of cointegration) between natural gas prices and each of the other I(1) variables using the Engle and Granger (1987) two-step procedure. The tests are first done with natural gas as the dependent variable in the cointegrating regression and then repeated with each of the other I(1) variables as the

---

[2]This is consistent with the evidence recently reported by Serletis and Andreadis (2004). In particular, they use daily observations on WTI crude oil prices at Chicago and Henry Hub natural gas prices at Louisiana (over the deregulated period of the 1990s) and various tests from statistics and dynamical systems theory to support a random fractal structure for North American energy markets.

TABLE 7.2

P-VALUES OF UNIT ROOT, COINTEGRATION, AND GRANGER CAUSALITY TESTS

| Variable | Unit root | | | Cointegration | | Granger causality | |
	WS	ADF	$Z(t_{\hat\alpha})$	x_t	y_t	$x_t \to y_t$	$y_t \to x_t$
Natural gas	.247	.263	.178				
Industrial production	.998	.947	.999	.433	.866	(3,1) .340	(3,3) .352
Consumer prices	.791	.371	.667	.161	.715	(5,5) .000	(4,12) .011
Crude oil	.284	.445	.603	.171	.463	(1,1) .205	(3,1) .270
Heating oil	.008	.026	.475	.171	.608	(11,2) .455	(1,1) .941
Propane	.370	.533	.447	.021	.334	(2,1) .767	(2,1) .084

Note: Results are reported using monthly data for the period January 1990 to March 2002.

x_t = Natural gas, y_t = (Industrial production, Consumer prices, Crude oil, Heating oil, Propane).

dependent variable. The results, under the 'Cointegration' columns of Table 7.2, suggest that the null hypothesis of no cointegration between natural gas prices and each of the other I(1) variables cannot be rejected (at the 5% level) in all cases.

Since we are not able to find evidence of cointegration, to avoid the spurious regression problem we test for Granger causality in the context of the following system

$$\Delta y_t = \alpha_1 + \sum_{j=1}^{r} \alpha_{11}(j)\Delta y_{t-j} + \sum_{j=1}^{s} \alpha_{12}(j)\Delta x_{t-j} + \varepsilon_{yt}, \qquad (7.1)$$

$$\Delta x_t = \alpha_2 + \sum_{j=1}^{r} \alpha_{21}(j)\Delta y_{t-j} + \sum_{j=1}^{s} \alpha_{22}(j)\Delta x_{t-j} + \varepsilon_{xt}, \qquad (7.2)$$

where $\alpha_1, \alpha_2, \alpha_{11}(j), \alpha_{12}(j), \alpha_{21}(j)$, and $\alpha_{22}(j)$ are all parameters and ε_{yt} and ε_{xt} are white noise disturbances. As in the previous section, we use x_t to denote logged natural gas prices and y_t to denote the logarithm of each of the other variables; since heating oil is a stationary series, its logged level is used in (7.1) and (7.2) instead of its logarithmic first difference.

In the context of (7.1) and (7.2) the causal relationship between y_t and x_t can be determined by first fitting equation (7.1) by ordinary least squares and obtaining the unrestricted sum of squared residuals, SSR_u. Then by running another regression equation under the null hypothesis that all the coefficients of the lagged values of Δx_t are zero, the restricted sum of squared residuals, SSR_r, is obtained. The statistic

$$\frac{(SSR_r - SSR_u)/s}{SSR_u/(T-1-r-s)},$$

has an asymptotic F-distribution with numerator degrees of freedom s and denominator degrees of freedom $(T-1-r-s)$. T is the number of observations, r represents the number of lags of Δy_t in equation (7.1), s represents the number of lags for Δx_t, and 1 is subtracted out to account for the constant term in equation (7.1).

If the null hypothesis cannot be rejected, then the conclusion is that the data do not show causality. If the null hypothesis is rejected, then the conclusion is that the data do show causality. The roles of y_t and x_t are reversed in another F-test [as in equation (7.2)] to see whether there is a feedback relationship among these series.

We used the AIC with a maximum value of 12 for each of r and s in (7.1) and (7.2) and by running 144 regressions for each bivariate relationship we chose the one that produced the smallest value for the AIC. We present these optimal lag length specifications in the last two columns of Table 7.2

together with p-values for Granger causality F-tests based on the optimal specifications. Clearly, there is evidence of a feedback relationship between natural gas prices and consumer prices (at about the 1% level). There is no evidence of industrial production causing natural gas prices, although in the previous section we established that natural gas prices are procyclical and lagging the cycle.

Finally, there is no evidence of a causal relationship between crude oil prices and natural gas prices. This is perhaps due to the fact that the Henry Hub natural gas market is much more segmented than the WTI crude oil market. For example, when crude oil prices change, they tend to change world-wide whereas the price of natural gas can easily change in North America without any changes in natural gas prices in other continents. This follows because transportation of natural gas by pipeline is far cheaper than transportation by ship (liquefied natural gas).

7.4 Conclusion

We have investigated the cyclical behavior of natural gas prices, using monthly data for the period that natural gas has been traded on organized exchanges and the methodology suggested by Kydland and Prescott (1990). Based on stationary Hodrick and Prescott (1980) and Baxter and King (1999) cyclical deviations, our results indicate that natural gas prices are procyclical and lag the cycle of industrial production. Moreover, natural gas prices are positively contemporaneously correlated with U.S. consumer prices and lead the cycle of consumer prices, raising the possibility that natural gas prices might be a useful guide for U.S. monetary policy, like crude oil prices are, possibly serving as an important indicator variable.

However, using lead-lag relationships to justify causality is tenuous. For this reason we also investigated the causality relationship between natural gas prices and U.S. industrial production and consumer prices, as well as between natural gas prices and each of crude oil, heating oil, and propane prices. This examination utilized state-of-the-art econometric methodology, using the single-equation approach. Our results indicate that industrial production does not Granger cause natural gas prices (although natural gas prices are procyclical and lag the cycle) and that there is a feedback relationship between natural gas prices and consumer prices.

Our results regarding the absence of a causal relationship between natural gas prices and crude oil prices are consistent with the evidence recently reported by Serletis and Rangel-Ruiz (2004) who investigate the strength of shared trends and shared cycles between WTI crude oil prices and Henry Hub natural gas prices using daily data from January, 1990 to April, 2001.

Based on recently suggested testing procedures they reject the null hypotheses of common and codependent cycles, suggesting that there has been 'de-coupling' of the prices of these two energy sources as a result of oil and gas deregulation in the United States.

Chapter 8

Futures Trading and the Storage of North American Natural Gas

*Apostolos Serletis and Asghar Shahmoradi**

8.1 Introduction

This chapter extends the work in Serletis and Shahmoradi (2005) by testing the theory of storage in the North American natural gas market. The theory of storage is the dominant model of commodity futures prices — see, for example, Brennan (1958), Telser (1958), and Working (1949). It postulates that the marginal convenience yield on inventory falls at a decreasing rate as aggregate inventory increases.

The hypothesis of the theory of storage can be tested in one of two ways — directly, by relating the convenience yield to inventory levels, or indirectly, as in Fama and French (1988), by testing its implications about the relative variation of spot and futures prices. Given the difficultly of defining and measuring the relevant inventory, we use the Fama and French (1988) indirect tests, based on the relative variation of spot and futures prices — see also Serletis and Hulleman (1994) for a similar approach.

The chapter is organized as follows. Section 8.2 discusses the theory of storage, Section 8.3 the data, and Section 8.4 presents the empirical results.

---

*Originally published in *OPEC Review* (2006), 19-26. Reprinted with permission.

Section 8.5 investigates the robustness of our results and the final section summarizes the chapter.

8.2 Testing the Theory of Storage

We use the Fama and French (1988) indirect test, based on the relative variation of spot and futures prices, to test the theory of storage in the natural gas market — see Serletis and Hulleman (1994) for similar tests in the crude oil, heating oil, and unleaded gas markets. The theory postulates that the marginal convenience yield on inventory falls at a decreasing rate as average inventory increases. This hypothesis can be tested either directly by relating the convenience yield to inventory levels, or indirectly by using the Fama and French (1988) test.

Following Fama and French (1988) and Serletis and Hulleman (1994), we consider the interest-adjusted-basis equation

$$\frac{F(t,T) - S(t)}{S(t)} - R(t,T) = \frac{W(t,T) - C(t,T)}{S(t)} \tag{8.1}$$

where $F(t,T)$ is the futures price at time t for delivery of the commodity at T, $S(t)$ is the spot price at t, $R(t,T)$ is the interest rate at which market participants can borrow or lend over a period starting at date t and ending at date T, $W(t,T)$ is the marginal warehousing cost, and $C(t,T)$ is the marginal convenience yield. According to this equation, the observed quantity on the left-hand side — the interest-adjusted basis — is the difference between the relative warehousing cost, $W(t,T)/S(t)$, and the relative convenience yield $C(t,T)/S(t)$.

Assuming that the marginal warehousing cost is roughly constant, that the marginal convenience yield declines at a decreasing rate with increases in inventory [see, for example, Brennan (1958) and Telser (1958)], and that variation in the marginal convenience yield dominates variation in the marginal warehousing cost, we can use the interest-adjusted-basis equation to develop testable hypotheses about the convenience yield. For example, when inventory is low the relative convenience yield is high, and larger than the relative warehousing cost, so the interest-adjusted basis becomes negative. On the other hand, when inventory is high the relative convenience yield falls toward zero, and the interest-adjusted basis becomes positive.

To test the theory of storage in the natural gas market, we use the Fama and French (1988) indirect test. In particular, using the sign of the interest-adjusted basis as a proxy for high (+) and low (−) inventory, the prediction of the theory that shocks produce more independent variation in spot and futures prices when inventory is low implies that the interest-adjusted basis is more variable when it is negative — see French (1986) for a derivation and detailed discussion. The indirect test of the theory of storage is the preferred approach over a direct test, due to the lack of available data for the convenience yield. As such, the investigation of the relationship between the marginal convenience yield and the price of the underlying asset, as established by the interest adjusted basis equation, is not explored empirically here.

8.3   The Data

We use daily data over the period from May 1, 1990 to July 12, 2002. In particular, we use 3-month, 6-month, and 1-year New York Mercantile Exchange (NYMEX) natural gas futures prices from Norman's Historical Data (http://www.normanshistoricaldata.com). We use the Henry Hub spot natural gas price, obtained from the Alberta Department of Energy. Moreover, we use daily 3-month, 6-month, and 1-year (U.S.) Treasury constant maturity interest rates (from http://www.federalreserve.gov/releases/h15/data.htm) to construct the corresponding interest-adjusted bases as an annualized rate of return.

8.4   Empirical Results

To test the prediction that the interest-adjusted basis is more variable when it is negative, we report in Panel A of Table 8.1 the number of positive, negative, and total observations of the interest-adjusted basis for 3-month, 6-month, and 1-year futures contracts. Panel B shows the average values of these interest-adjusted bases, and Panel C reports the standard deviations of changes in the interest-adjusted bases. Clearly, the standard deviation is larger when the interest-adjusted basis is negative than when it is positive for all three futures contracts, thereby providing evidence that the natural gas market passes the Fama and French indirect test.

The theory of storage also predicts that supply and demand shocks cause approximately equal changes in spot and futures prices when inventory levels are high (positive interest-adjusted basis), but cause spot prices to change more than futures prices when inventory levels are low (negative interest-adjusted basis). In order to test this, we report in Table 8.2 the ratio of the standard deviations of percent futures price changes to the standard deviations of percent spot price changes and compare these ratios across the positive and negative interest-adjusted bases samples. Clearly, the ratios are lower when the interest-adjusted basis is negative than when it is positive, thereby confirming the theory of storage prediction about the response of spot and futures prices to shocks.

TABLE 8.1

Summary Statistics for Daily 3-Month, 6-Month, and 1-Year Natural Gas Interest-Adjusted Bases

Basis	Positive	Negative	All
	A. Number of Observations		
3-Month	1482	1533	3015
6-Month	1512	1503	3015
1-Year	1198	1817	3015
	B. Average Values		
3-Month	0.505	-0.644	-0.079
6-Month	0.344	-0.435	-0.044
1-Year	0.172	-0.206	-0.056
	C. Standard Deviations of Changes		
3-Month	0.621	0.675*	0.867
6-Month	0.287	0.350*	0.504
1-Year	0.163	0.171*	0.250

Notes: Sample period, daily observations: May 1, 1990 to July 12, 2002 (3015 daily obervations). Numbers are for observations when the interest-adjusted basis is positive (Positive), observations when it is negative (Negative), and for all observations (All). An asterisk indicates rejection of the null hypothesis of equal variances at the 5% level.

The final prediction of the theory of storage is that supply and demand shocks cause larger changes in near term futures as opposed to longer term futures. To test this, we once again look at the ratios in Table 8.2. For this prediction to hold, these ratios must fall as maturity dates increase. In this case, we do not wish to divide the sample based on inventory levels, and we therefore focus on the third column. We see that this prediction holds, and that shocks do cause greater variation in near-term futures than in long-term futures.

TABLE 8.2
RATIOS OF THE STANDARD DEVIATION OF PERCENT
FUTURES PRICE CHANGES TO THE STANDARD
DEVIATION OF PERCENT SPOT-PRICE CHANGES

Basis	Positive	Negative	All
3-Month	.653	.764*	.719
6-Month	.544	.452*	.496
1-Year	.466	.399*	.428

Note: Results are reported using daily data for the period May 1, 1990 to
July 12, 2002. An asterisk indicates rejection of the null hypothesis of equal
ratios at the 5% level.

8.5 Robustness

Although the T-bill rate is routinely used for calculations of the interest-adjusted basis, here we also investigate the robustness of our results to the use of alternative interest rate measures. In particular, we use 3- and 6-month Eurodollar rates (from http://www.federalreserve.gov/releases/h15/data.htm) to calculate the 3- month and 6-month interest-adjusted bases and report summary statistics in Tables 8.3 and 8.4 in the same way as those in Tables 8.1 and 8.2 based on the T-bill rates; 1-year Eurodollar rates are not available and this is why we do not report results for the 1-year Eurodollar-adjusted basis. The evidence in Tables 8.3 and 8.4 is consistent with the evidence in Tables 8.1 and 8.2, suggesting that our results regarding the predictions of the theory of storage are robust to the use of different interest rates in calculating the interest-adjusted basis for natural gas.

8.6 Conclusion

We tested the theory of storage in North American natural gas markets, using the Fama and French (1988) indirect test. This test of the theory of storage is the preferred approach over a direct test, due to the lack of available data for the convenience yield. We tested the prediction of the theory that, when inventory is high, large inventory responses to shocks imply roughly equal changes in spot and futures prices, whereas when inventory is low, smaller inventory responses to shocks imply larger changes in spot prices than in futures prices.

Our tests on spot and futures North American natural gas prices confirm these predictions of the theory of storage.

TABLE 8.3

SUMMARY STATISTICS FOR DAILY 3-MONTH AND
6-MONTH NATURAL GAS EURODOLLAR-ADJUSTED BASES

Basis	Positive	Negative	All
	A. Number of Observations		
3-Month	1475	1540	3015
6-Month	1504	1511	3015
	B. Average Values		
3-Month	0.505	-0.644	-0.082
6-Month	0.344	-0.436	-0.047
	C. Standard Deviations of Changes		
3-Month	0.622	0.676*	0.867
6-Month	0.287	0.351*	0.505

Notes: Sample period, daily observations: May 1, 1990 to July 12, 2002
(3015 daily obervations). Numbers are for observations when the interest-
adjusted basis is positive (Positive), observations when it is negative
(Negative), and for all observations (All). An asterisk indicates rejection
of the null hypothesis of equal variances at the 5% level.

TABLE 8.4

RATIOS OF THE STANDARD DEVIATION OF PERCENT
FUTURES PRICE CHANGES TO THE STANDARD
DEVIATION OF PERCENT SPOT-PRICE CHANGES

Basis	Positive	Negative	All
3-Month	.664	.759*	.719
6-Month	.542	.454*	.496

Notes: Results are reported using daily data for the period May 1, 1990 to
July 12, 2002. An asterisk indicates rejection of the null hypothesis of equal
ratios at the 5% level.

Part 3

Electricity Markets

Overview of Part 3

Apostolos Serletis

The following table contains a brief summary of the contents of the chapters in Part 3 of the book. This part of the book consists of three chapters (two of which have not been previously published) devoted to electricity issues in Alberta's deregulated electricity market.

Electricity Markets

Chapter Number	Chapter Title	Contents
9	Power Trade on the Alberta-BC Interconnection	Chapter 9 assesses the amount of power trade across the Alberta-BC interconnection and focuses on the fundamental role played by cross-border trade of electricity in restructured wholesale power markets.
10	Imports, Exports, and Prices in Alberta's Deregulated Power Market	It provides a study of the relationship between electricity prices and imports and exports for peak hours, off-peak hours, and all hours, using data over the period from January 1, 2000 to July 31, 2005 from Alberta's (deregulated) spot power market.
11	Cointegration Analysis of Power Prices in the Western North American Markets	This chapter empirically investigates the extent of integration in the main Western North American power markets and to what extent deregulation and open access to transmission policies have removed barriers to trade among them.

Chapter 9:

In this (previously unpublished) chapter we focus on the fundamental role played by cross-border trade of electricity in restructured wholesale power markets. First, we describe the economic and physical implications of engaging in an inter-systemic exchange of energy. Then, we assess the amount of power trade across the Alberta-BC interconnection and we find that the creation of a single regional transmission organization that operates the transmission grids in the Western region would increase trading opportunities and increase the efficiency in the utilization of interconnectors.

Chapter 10:

This chapter investigates whether Alberta's power interconnection lines can become a tool of market power abuse. In doing so, it tests for Granger causality from imports and exports to pool prices, using data on prices, load, imports, and exports for peak hours, off-peak hours, and all hours (over the period from January 1, 2000 to July 31, 2005). Interpreting causality in terms of predictability, it rejects the null of no causality from imports and exports to the pool price, thereby providing evidence for potential market power abuse in Alberta's power market.

Chapter 11:

This (previously unpublished) chapter aims to determine the extent of market integration in the main Western North American power markets and to what extent deregulation and open access to transmission policies have removed barriers to trade among them. In doing so, it tests for cointegration between power prices; it develops an error correction model; and then looks for causal relationships between the price dynamics in the Alberta, Mid Columbia, California Oregon Border, California NP15, and California SP15 power markets.

Chapter 9

Power Trade on the Alberta-BC Interconnection

Mattia Bianchi and Apostolos Serletis

9.1 Introduction

The functioning of an electric system is subject to specific physical laws that apply to electricity. Differently from other commodities, electricity cannot be stored and power supply and demand must be continuously balanced. Its instantaneous nature creates complexities which have to be managed through engineering operating practices. As a consequence, a very tight interaction interdependence exists between technical characteristics of electricity and market structures. The laws of physics dictate certain essential attributes of market operations, while market design distortions may become sources of technical constraints and dysfunctions.

Physical characteristics of power have significant consequences on the organization of the electric industry, which has traditionally been regulated. Recently, a general trend toward deregulation of electric industry has been under way in several countries around the world. The process of restructuring the North American electricity industry began in the early 1990s and promoted the development of electricity spot markets. The Federal Energy Regulatory Commission (FERC) deregulated wholesale electricity markets in the United States under the 1992 Energy Policy Act, obliging transmission-owning utilities to open their transmission systems to market participants. FERC was given authority to guarantee openness and fairness in regional power markets and transmission systems. Order 888 and Order 889, issued by FERC in 1996, aimed to provide all market participants an equal transmission service and direct access to real-time information on tariffs and available transfer capacity. The availability of unbundled transmission service has enhanced the exploitation of the Western Interconnection grid, resulting in increased opportunities for power trade among different regions.

Restructured wholesale markets are intended to achieve economic efficiency by capturing the gains from the trade of electricity among many market players. Successful competitive markets work through the interaction of private, decentralized trading and investment decisions to minimize the total cost of electricity. Competition puts a downward pressure on the profit margins of generators and suppliers and provides an incentive to reduce costs. Independent System Operators (ISOs) play a key role in coordinating the dispatch of electricity supply to meet the demand, so that power is supplied at the lowest cost possible. Better investment decisions and innovations can be expected from competitive market participants since they assume the risks of their investments.

In this chapter we focus on the fundamental role played by cross-border trade of electricity in restructured wholesale power markets. First, we describe the economic and physical implications of engaging in a intersystemic exchange of energy. Then, we assess the amount of power trade across the Alberta-BC Interconnection and we find that the creation of a single regional transmission organization that operates the transmission grids in the Western region would increase trading opportunities and increase the efficiency in the utilization of interconnectors.

9.2   Wholesale Trade of Electricity: Economic and Physical Implications

The necessary condition for separate electric systems to exchange power is the existence and operation of interconnectors — transmission lines connecting different control areas.[1] Inter-ties are an essential part of a fair, open and competitive market since they consent to import and export energy whenever profit opportunities arise or security problems exist. If properly operated, tie-lines contribute to minimizing the costs of supplying energy and maximizing the total surplus, both to consumers and to producers, by ensuring a better allocation of resources. Interconnectors can act

---

[1]The words "interconnector," "inter-tie," and "tie-line" are used interchangeably.

as both substitute for and a complement to generation. They represent a tool to ensure security and solve emergency conditions by sharing reserves in case of failures. In addition, they provide access to external low-priced markets thus diversifying the mix of power sources on which a system relies. Inter-ties improve the efficient operation of power systems by allowing the economic trade of electricity between neighbouring regions; according to existing market conditions, traders buy and sell energy arbitraging price differentials.

Interconnectors allow two separate electric systems to exchange and trade power. Cross-border trade of electricity results in importing and exporting activities by market participants. Such interchange transactions involve the purchase of power in, say, market A; the purchase of transmission service on the grids that are crossed; the sale of power in, say, market B. Since traders are opportunistic and attempt to arbitrage the difference between the prices in the two markets, electricity generally will flow from low priced areas to high priced areas. Generators and marketers export power profiting by selling it at prices above their marginal costs, while utilities import power from cheaper sources thus reducing the cost of supply. Transactions take place until the gains from trade are eliminated.

Trade activity tends to reduce the price difference between markets. In the case of an interconnector whose capacity does not limit the exchange of power, the price difference between the interconnected markets would be equal to the transportation and transaction costs. However, this is hardly ever true in the real electricity industry. Generally, physical and reliability constraints limit cross-border exchanges and transmission congestion keeps energy prices different.

In order to wheel energy across separate areas, market participants demand transmission services. Since the transfer capacity of an interconnector is a scarce resource, a transmission rate is charged by the system operator to any MWh of power flowing. Such charges represent the transportation cost to move electricity over a network of copper or aluminium wires from the injection to the withdrawal node. The rates are generally fixed in advance in order to influence participants' behaviour appropriately. Charges, which are paid by traders who engaged in the transaction, limit the demand for transmission services as they reduce the profitability of a cross-border trade.

Similarly, power losses lower the potential gains from a trade of electricity. Losses occur during transmission and arise from the specific physical nature of electricity. Energy transferred over transmission wires is lost as heat, proportionally to the square of electricity flows (Joule Effect). A loss factor, which is usually expressed as a percentage, is computed by the system operator. The value of the lost power depends on the value of electricity. Therefore, power traders undertaking an inter-systemic transaction have to consider the impact of transmission charges and power losses which lower the profit from the trade. In order for the transaction to be profitable, price differential must exceed losses and wheeling charges.

9.3 The Alberta-British Columbia Interconnection

We now consider a scenario involving trade of electricity between the Alberta's electric system and the Mid-Columbia power market, which is the trading hub most commonly referenced in the Pacific North West. In order to be transferred across these areas, power must flow on the Alberta-British Columbia Interconnection. Currently, transmission constraints and complications in market procedures and rules used to manage the Alberta-BC Interconnection limit cross-border trade of electricity.

Physical transmission capacity acts as a bottleneck. Alberta is a peripheral market, scarcely interconnected to the Pacific power markets. The interconnection capacity (about 800 MW) as a percentage of peak load is lower in Alberta than in any other province in Canada: approximately 12% in Alberta compared to 40% in British Columbia. Also, Alberta has the lowest import/export capacity among the major systems in North America.

Furthermore, market imperfections render arrangements for power exchanges outside Alberta very complicated. First, Alberta and the neighbouring regions have very different tariff regimes and this constitutes a significant obstacle to marketers. For instance, while in British Columbia transfer capacity can be reserved from hours ahead to years ahead of the actual energy flow, on the Alberta portion of the interconnector it has to be acquired on a day-by-day basis. Also, the transfer capacity is allocated on a first-come-first-served principal. This method, used to avoid congestion, generates inefficient solutions since the interconnector is not used by the transactions that have the higher economic value, but by those transactions that were quicker in submitting the reservation.

However, the main market barrier that limits the efficient use of the Interconnector is the aggregation of transmission charges. Currently, electricity crossing states and regions may pass over grids controlled by several utilities in order to be delivered to customers. For each transmission system crossed, a rate is charged by each utility, independent of the distance between the injection and the withdrawal node. The accumulation of multiple rates is called "rate pancaking". This condition discourages long-distance transactions and renders trade uneconomic. Rate pancaking applies to cross-border exchanges between Alberta and Mid-C markets. A market participant wishing to sell into Alberta power purchased in Mid-C has to pay transmission charges to the Bonneville Power Administration (BPA), which controls over 75% of the grid in the Pacific Area, to British Columbia Transmission Corporation (BCTC) and to the Alberta Electric System Operator (AESO). This reduces the profitability of trades and the efficient use of the interconnector.

9.4 Empirical Analysis

Now we want to analyze cross-border trading activity between the Alberta and Mid Columbia power markets. For the purpose of our analysis, we use data on power prices in Alberta and Mid Columbia; on the actual power flows across the Alberta-BC Interconnection and on the available transfer capacity (ATC) on the Interconnection; on transmission charges and on loss factors existing in BPA, BCTC, and AESO electric systems. These data consist of two daily observations: peak and off-peak. The sample period is from January 1st, 2000 to September 30th, 2005.

We assume that tie line users are motivated to export or import in order to arbitrage the two markets. When price differentials between Alberta and Mid-Columbia markets exceed transmission charges and losses, a trading opportunity exists. To the extent that the available transfer capacity (ATC) on the Alberta-BC Inter-tie is exploited by exchanging power under such price differentials, the opportunity is seized and profit from the trade is equal to

$$\Pi_{\text{gained}} = (P_s - P_{p-\text{adj}}) \times Q$$

where P_s represents the sale price of power, $P_{p-\text{adj}}$ the purchase price (adjusted for losses and transmission charges), and Q is the power flow exchanged across the interconnector.

To the extent that the available transfer capacity (ATC) is not utilized, no profit arises and the opportunity is missed. The total opportunity is the sum of the seized opportunity and the missed opportunity and is equal to:

$$\Pi_{\text{total}} = (P_s - P_{p-\text{adj}}) \times \text{ATC}$$

$$= \Pi_{\text{gained}} + \Pi_{\text{missed}}$$

The efficiency indicator (EI) measures the gained opportunity as a percentage of the total opportunity:

$$EI = \frac{\Pi_{\text{gained}}}{\Pi_{\text{total}}}$$

According to our analysis for peak hours, for 53% of the period, price differentials (adjusted for losses and transmission charges) were not large enough to offer a trading opportunity. For the remaining 47% of the period, trading opportunities exist to be seized. An import transaction of electricity from Mid-C to Alberta was profitable for the 28% of the period, while an export from Alberta to Mid-C was profitable for 19% of the period.

With regard to off-peak hours, for 61% of the period of analysis, price differentials (adjusted for losses and transmission charges) were not large enough to offer a trading opportunity. For the remaining 39% of the period, trading opportunities exist to be seized. An import transaction of electricity from Mid-C to Alberta was profitable for 20% of the period, while an export from Alberta to Mid-C was profitable for 19% of the period.

Now we focus on the observations when trading opportunities occurred. Together with the price differentials, we consider the actual volumes of imports and exports exchanged across the interconnector and the available transfer capacity. In doing so, we estimate the gained and total opportunity from trade and we get a value for the efficiency indicator on a day by day basis. Over the period from January 2000 to September 2005, under the assumptions made, the total profit earned by power traders by using the Alberta-BC Interconnection is about $188 million. The majority of this comes from export transactions (96%) from Alberta to Mid-C. The reason for that are the extremely high gains realized in 2000 and in 2001, primarily through exporting low priced energy from Alberta and selling it into Pacific U.S. markets, which were experiencing soaring price spikes. The total profit that could have been potentially earned is about $341 million.

Over the whole period of analysis, the overall efficiency indicator for peak hours, calculated as average of all the daily efficiency indicators, shows an efficiency of 39% for imports and of 70% for exports. The same indicator for off-peak hours shows an efficiency of 11% for imports and of 85% for exports. Thus, the efficiency indicator for imports is significantly lower than the same indicator for exports. This causes the profits gained from importing activities (4% of the total) to be extremely little if compared to ones from exports. In fact, if we set an identical efficiency indicator for imports and exports by considering aggregate total opportunities by trades over the period 2000-2005 (thus we take an EI equal to 100%), we see that the potential profits arising from power imports to Alberta would be about 43% of the total $341 million, much higher than the previous 4%.

Therefore, a first conclusion is that export and import transactions roughly present the same potential profitability. Nevertheless, in the last five years, the earned profits from power trade came almost entirely from exporting activity in 2000 and 2001. Two are the main reasons for that. The first is that the available transfer capacity for imports is much less exploited than the corresponding for exports. This can be due to the fact that the transfer capacity made available for flows of imported energy includes the capacity reserved for Remedial Action Scheme (RAS), which is reserved for reliability purposes and cannot be utilized for trades of electricity. As a consequence, the unused capacity is larger. Moreover, market mechanisms and procedures between Alberta and British Columbia may render the arrangement of importing transactions more difficult than of exporting transactions. Finally, the average ATC for exports (209MW during peak hours and 479MW during off-peak hours) is lower than the ATC for imports (581MW during peak hours and 552MW during off-peak hours). As a consequence, the Alberta-BC Interconnection is more often congested in the East-West direction and this pushes the exports' efficiency indicator up.

The second main reason for the extremely low gained profits from import transactions comes from price differentials. Over the period from January 2000 to September 2005, the average price differential faced by exporters was three times larger than the average price differential accruing to importers, both for peak and off-peak hours. Again, the extraordinary market conditions in 2000 and 2001, following the major electricity crisis in California, are responsible for this result. Exceptionally high prices in Pacific U.S. markets cause exports of electricity from Alberta to be very profitable. In 2000 and 2001, power traders managed to earn large profits, which did not occur the following years. In fact, if we limit the analysis over the time horizon 2002-2005, we observe that the gain from an import trade is larger than the gain from an export, meaning that importing one MWh of electricity to Alberta is on average more profitable than exporting it. However, the price differences are close to each other and are far lower than the ones experienced in 2000 and 2001.

In conclusion, results from the analysis shows that for more than half of the period, market conditions did not support trade of electricity. Therefore, there is room for improvements in the utilization of the Alberta-BC Inter-tie for trading purposes. Different tariff regimes constitute a significant obstacle to marketers. In fact, in the next section, we find evidence that the introduction of a regional transmission organization in the Pacific area increases trading opportunities between markets and fosters the efficient utilization of the Interconnection.

9.5   An RTO Scenario in the Western Region

The Regional Transmission Operator (RTO) initiative represents the biggest step by FERC to create seamless and non-discriminatory open access to transmission lines that were previously controlled in terms of access and pricing by vertically integrated utilities. According to FERC, open access transmission is the foundation to competitive wholesale power markets in North America. This vision entails the creation of a single system operator and the development of a single seamless market, as opposed to the current independent transmission systems with numerous control areas, transmission owners and business practices. Following the introduction of RTOs, the existing transmission systems that are generally designed to move power within local utility systems, would increasingly be used to enable power sales across large geographic areas.

FERC has always considered the West as one "natural market" for electricity. Moreover, FERC has stated in different orders that it welcomes Canadian participation in RTOs since electricity markets are North American in nature. In the light of this, we consider a scenario where the Alberta, British Columbia and Mid-C power markets are operated as a single regional market. In this case, the interconnectors are treated as normal transmission lines and cross-border flows of electricity are scheduled by a single system operator. Since an RTO is regional, it assures the efficient delivery of power over long distance within its region by removing pancaked rates and providing harmonized market mechanisms and procedures.

Using the data from the previous analysis, we model the existence of an RTO by removing the pancaked rates from the BPA, BCTC, and AESO transmission systems. Thus we assume no fixed charges by setting the three transmission charges equal to $0/MWh. In this case an opportunity of profit would exist if the price differentials simply exceeded the costs of power losses. Then we evaluate the trading opportunities arising from the exchange of electricity across the Alberta-BC Interconnection. We find that the removal of fixed charges in a RTO scenario would appear to encourage additional trade 24% of the time during peak periods and 28% of time during off-peak periods.

Then, we focus on the observations when trading opportunities would occur. Together with the price differentials, we consider the available transfer capacity on the Interconnector. We do not consider the data about imports and exports that occurred over the period 2000-2005. The reason for it is that the actual flows of electricity moving across the Interconnector largely depend on price differentials between markets. Since in an RTO scenario price differentials have changed due to the removal of transmission charges, the volumes of imports and exports are not meaningful under these price differentials and thus they are not included in this scenario.

Nevertheless, we can correctly assume that in an RTO scenario, they would be equal to the available transfer capacity. In fact, if Alberta, BC, and Mid-C were operated as a single regional market and we assume locational marginal pricing as a method for congestion management, the price differentials between them would not exceed the transmission loss differentials as long as Inter-tie capacity is available and not being fully used. Therefore, when the price differentials exceed transmission loss differentials, that is when a trading opportunity exists, the Inter-tie is congested, the actual flows of electricity on it equalling the available transfer capacity. Thus, in an RTO scenario, the efficiency indicator, when trading opportunities exist, is equal to 100% because the interconnector is fully utilized and the gained opportunity is equal to the total opportunity.

According to the results, the average gain from trading activity, that is the profit that could be gained by the sale of 1 MWh of electricity, is lower in an RTO scenario, compared to the real situation of Alberta, BC, and Mid-C being separate control areas. This is true both for imports and exports during peak and off-peak hours. Such a result is due to the higher number of trading opportunities when price differentials are very small, resulting from the removal of transmission charges. Similarly, the average total opportunity is smaller, since the ATC is almost unchanged in the two scenarios. However, due to the huge increase in the number of profitable trading opportunities and the 100% efficiency indicator in an RTO scenario, the aggregate gained opportunity, which represents the amount of profits that would arise from trading activities over the period 2000-2005, is much larger, $430 million compared to $188 million. Following the introduction of a regional transmission organization in Western North America, there would be additional $242 million worth of profits arising from augmented trade of electricity. This corresponds to a 129% increase from the real situation.

9.6 Conclusion

Cross-border trade of electricity plays a fundamental role in restructured wholesale markets. The existence and exploitation of interconnectors consent power markets to act as open systems. Inter-systemic exchange of power represents a potential source of efficiency and reliability. However, transmission constraints and market imperfections limit the optimal use of inter-ties.

Importing and exporting activities by market participants are already a reasonably big business but need to grow in order for the benefits to be larger and more fairly distributed. In fact, the profits from interregional power trade, instead of adding up to power traders, can be considered as gains in consumer and producer surpluses. They can be thought as savings in production costs accruing to consumers of electricity, due to access to cheaper resources.

A viable solution to foster cross-border exchange of electricity is the creation of regional transmission organizations. RTOs reap the gains from energy trade over large regions by removing pancaked transmission rates applied by each system operator, by promoting more efficient dispatching schedule and supporting the reliability of the grid. The existence of RTOs guarantees the optimal dispatch of generation and the efficient utilization of interconnector, which minimize the total cost of energy. Due to the particular physical characteristics of electricity, coordination and openness ensure the efficient and reliable operation of power systems.

Chapter 10

Imports, Exports, and Prices in Alberta's Deregulated Power Market

*Apostolos Serletis and Paul Dormaar**

10.1 Introduction

Recent leading-edge research has applied various innovative methods for modeling spot wholesale electricity prices — see, for example, Bunn (2004), Deng and Jiang (2004), León and Rubia (2004), Serletis and Andreadis (2004), Czamanski *et al.* (2006), and Hinich and Serletis (2006). These works are interesting and attractive, but have taken a univariate time series approach to the analysis of electricity prices. From an economic perspective, however, the interest in the price of electricity is in its relationship with the prices of various underlying primary fuel commodities [see, for example, Serletis and Shahmoradi (2006)] as well as in its relationship with import and export activity.

As the Market Surveillance Administrator (2005) of the Alberta electric system recently put it

---

*Originally published In W. David Walls (ed.) *Quantitative Analysis of the Alberta Power Market*. Van Horne Institute (2006). Reprinted with permission.

"the role and influence of imports and exports into/out of Alberta via the BC interconnection has long been a contentious issue amongst industry stakeholders. A recent issue that has been expressed by some participants concerns the occurrence of imports that appeared to be unprofitable based on economics using the appropriate market index prices. The concern was not so much that the observed imports were unprofitable, but rather that the motivation behind the import behaviour was a desire to influence Pool prices — in this case, to push Pool prices down."

Investigating the relationship between power prices and power imports and exports in the Alberta spot power market is our primary objective in this paper. In doing so, we use data on prices, load, imports, and exports for peak hours, off-peak hours, and all hours (over the period from January 1, 2000 to July 31, 2005) and test for causal relationships from power imports and power exports to power prices.

The paper is organized as follows. Section 10.2 discusses the role of imports and exports and their potential effect on power prices. Section 10.3 is devoted to data issues, Section 10.4 presents the causality model, and Section 10.5 the results of our empirical analysis. The final section briefly concludes the paper.

10.2   The Role of Imports and Exports

In deregulated power marketplaces, competitive market forces — the laws of supply and demand — guide electricity price formation. Factors affecting demand, like temperature and time of day, and factors affecting supply, like natural gas prices and unit outages, determine the pool price. Imports and exports of electricity contribute to narrow the price arbitrage between different markets and help diversify the mix of electric power sources. For example, a region relying entirely on thermal conversion of fossil fuels, may import hydro-powered electricity, thus becoming less dependent on natural gas price volatility.

Imports and exports, however, being components of supply and demand, can potentially influence power prices in deregulated markets. In particular, exports act as additional demand. As demand increases, more expensive generation must be dispatched to serve load causing electricity prices to rise. Thus, all the factors that influence load in the importing region (temperature, lighting, etc.) influence power prices in the exporting market. On the other hand, imports act as additional supply. If cheap imports are available, more expensive generating units can be dispatched off causing electricity prices to fall. Of course, import capability depends on the available transmission capacity and on the situation in the exporting market. For example, if transmission congestion happens, importers who could supply at lower prices may not be able to move their power. Also, plant failures, maintenance outages, and weather conditions in the exporting market may reduce electricity supply — as, for example, little precipitation can make low priced hydro generation unavailable.

The power markets in Alberta and British Columbia are connected through an 800 MW tie line — see Table 10.1. This available transmission capacity (including the 150 MW tie line into Saskatchewan) represents about 11% of the Alberta peak demand. With regard to electricity power sources, British Columbia relies almost entirely on their rich hydro power resources (94%) whereas Alberta relies more on conventional thermal and combustion turbines, given its large coal and natural gas reserves. Also the electricity market design is very different between the two provinces. Alberta is the first Canadian wholesale and retail competitive marketplace whereas British Columbia is a single buyer market, where the single buyer purchases a planned amount of power from competing independent power companies.

The tie-line between Alberta and British Columbia works as a very large generating unit, supplying power energy to Alberta. In fact, being larger than any generating plant in Alberta, it has the potential to strongly affect the pool price in Alberta. On the demand side, players in British Columbia import power at low price times (evenings), thus sustaining off-peak prices. Then they spill water for export at the high priced times. Having excess capacity, British Columbia companies are able to export electricity not only into Alberta but also into California. For example, when the Alberta market is depressed, exporters in British Columbia can sell power into the California market and vice versa. These are common sales tactics pursued by power traders in British Columbia, suggesting that prices in one region will reflect the opportunity cost of selling into the other region.

On the regulatory side, AESO (Alberta electric system operator) implemented in late 2000 the Pool Price Deficiency regulation that disallowed exports and imports from setting the pool price. However, imports received an uplift payment if they were accepted at a price higher than the pool price. Late in 2001, the rules were changed such that imports and exports were price takers. However, importers and exporters implement strategies to manage their portfolios in response to market circumstances aiming to maximize their profit, and uneconomic imports by traders, potentially abusing their market power, have generated complaints from several stakeholders in Alberta. It has been argued that pool prices are no longer a function of market drivers, but depend on the opportunistic behaviour of a few market players whose intentional and repeated actions affect the pool price and weaken market confidence.

TABLE 10.1
ALBERTA GENERATING CAPACITY (MW)

Local generation	
Coal	5,840
Natural gas	4,903
Hydro	900
Wind	270
Biomass	178
Fuel oil	8
Subtotal	12,099
Interconnections	
British Columbia	800
Saskatchewan	150
Subtotal	950
Grand total	13,049

Source: Alberta Department of Energy.

In what follows we provide a preliminary investigation of whether power interconnection lines can become a tool of market power abuse, in which case they will have devastating effects on the competition and fairness of deregulated electricity markets. In doing so, we test for Granger causality from imports and exports to pool prices, interpeting causality merely in terms of predictability.

10.3   Data

We use hourly data on power (volume weighted) prices, load, imports, and exports, over the period from January 1, 2000 to July 31, 2005, from the AESO web site, at http://www.aeso.ca. In doing so, we make a distinction between peak hours, off-peak hours, and all hours. For the purposes of this study, peak hours includes weekday hours from 7:00 a.m. to 11:00 p.m. inclusive while off-peak hours includes weekday hours from 11:00 p.m. to 7:00 a.m. inclusive as well as all day Sunday. All hours include all hours throughout each day of the week. It is to be noted that the price series are volume weighted to give the price of hours with greater volume more weight than hours with less volume.

Figures 10.1 to 10.6 show the price, load, and net imports for all hours, peak hours, and off-peak hours and Table 10.2 provides some summary statistics for each series. Figures 10.7-10.8 show average hourly prices, load, and net imports for each day of the week. As can be seen from Figure 10.7 and Table 10.2, during peak hours Alberta is a net importer of electricity while during off-peak hours is a net exporter of electricity.

TABLE 10.2
SUMMARY STATISTICS

Statistic	Price	Load	Imports	Exports
All hours (48,936 observations)				
Mean	70.51	6,824	143	138
Standard error	90.40	779	156	182
Skewness	4.16	0.02	1.60	1.20
Kurtosis	22.22	-0.30	2.41	0.29
Peak hours (27,952 observations)				
Mean	87.87	7,159	188	58
Standard error	106.42	679	171	108
Skeweness	3.53	-0.03	1.22	2.30
Kurtosis	15.07	-0.73	0.99	5.04
Off Peak hours (20, 984 observations)				
Mean	47.39	6,379	83	244
Standard error	55.11	672	106	205
Skewness	5.54	0.04	2.37	0.33
Kurtosis	50.77	-0.45	7.68	-1.10

10.4   Granger Causality Tests

We test for Granger causality from imports and exports to power prices using the levels of the variables (since according to Figures 10.1-10.6 they appear to be stationary) in the context of the following model

$$\text{Price}_t = a_0 + a_1 t + \sum_{j=1}^{r} \alpha_j \text{Price}_{t-j} + \sum_{j=1}^{s} \beta_j \text{Imports}_{t-j}$$

$$+ \sum_{j=1}^{q} \gamma_j \text{Exports}_{t-j} + \sum_{j=1}^{p} \delta_j \text{Load}_{t-j} + \varepsilon_t \tag{10.1}$$

where a_0, a_1, α_j, β_j, γ_j, and δ_j are all parameters, t is a time trend, and ε_t is a white noise disturbance. Note that in doing so, we control for the effects that are due to movements in load (defined to exclude imports and exports).

In the context of (10.1), causality from (say) Imports to Price can be determined by first fitting equation (10.1) by ordinary least squares and obtaining the unrestricted sum of squared residuals, SSR_u. Then by running another regression equation under the null hypothesis that all the coefficients of the lagged values of Imports$_t$ are zero, the restricted sum of squared residuals, SSR_r, is obtained. The statistic

$$\frac{(SSR_r - SSR_u)/s}{SSR_u/(T - r - s - q - p - 2)},$$

has an asymptotic F-distribution with numerator degrees of freedom s and denominator degrees of freedom $(T - r - s - q - p - 2)$. T is the number of observations, r represents the number of lags of Price$_t$ in equation (10.1), s represents the number of lags for Imports$_t$, q the number of lags of Exports$_t$, p the number of lags of Load$_t$, and 2 is subtracted out to account for the constant term and the trend in equation (10.1).

If the null hypothesis cannot be rejected, than the conclusion is that the data do not show causality. If the null hypothesis is rejected, then the conclusion is that the data do show causality.

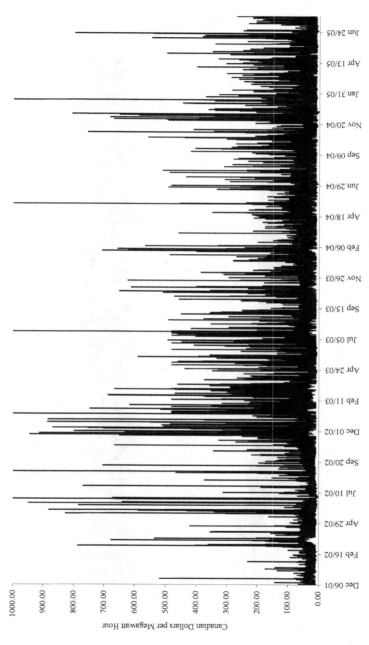

Figure 10.1: Alberta Power Prices: All Hours

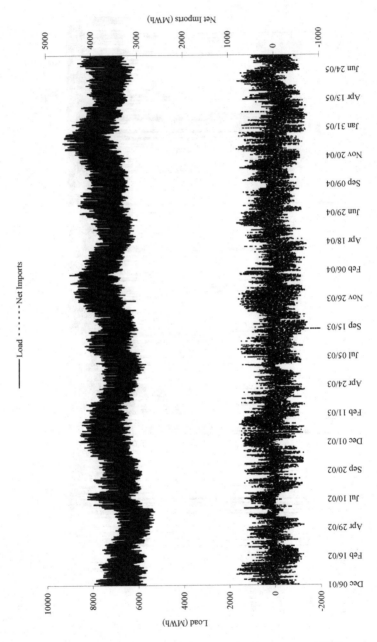

Figure 10.2: Alberta Power Load and Net Imports: All Hours

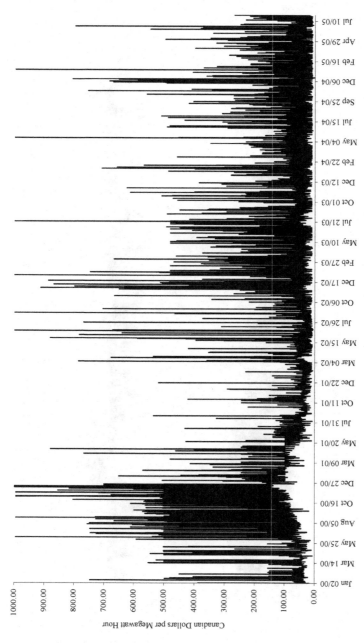

Figure 10.3: Alberta Power Prices: Peak Hours

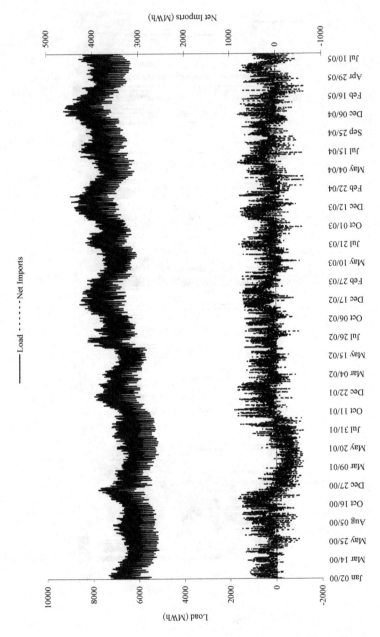

Figure 10.4: Alberta Power Load and Net Imports: Peak Hours

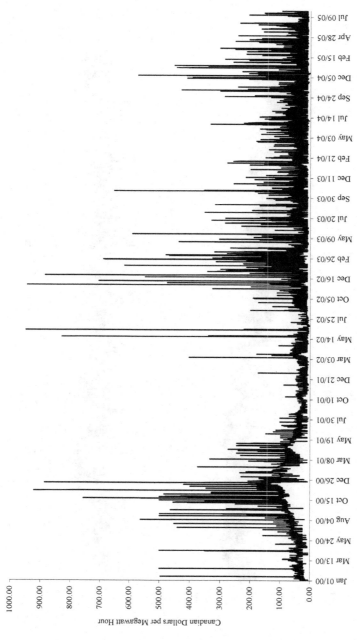

Figure 10.5: Alberta Power Prices: Off Peak Hours

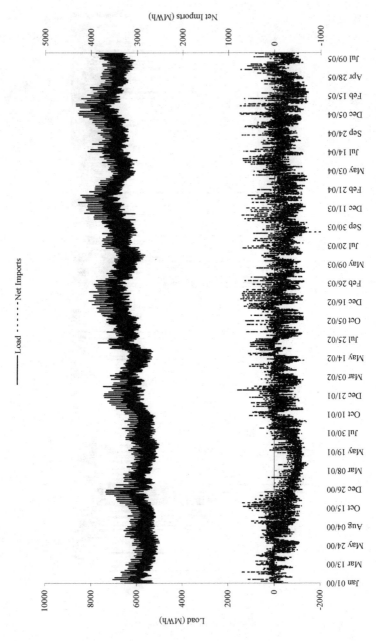

Figure 10.6: Alberta Power Load and Net Imports: Off Peak Hours

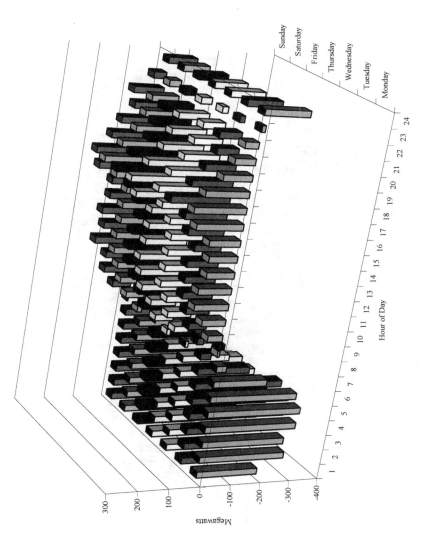

Figure 10.7: Alberta Power, Average Net Imports

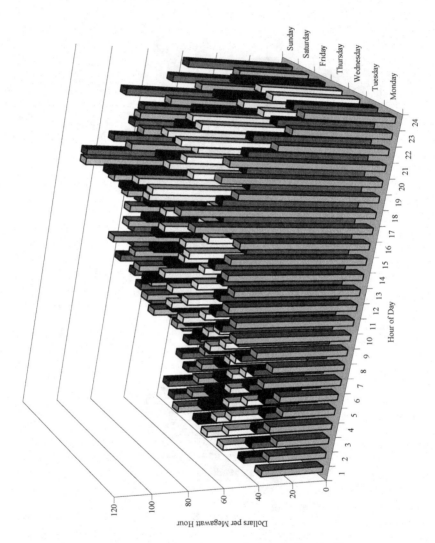

Figure 10.8: Alberta Power, Average Hourly Price

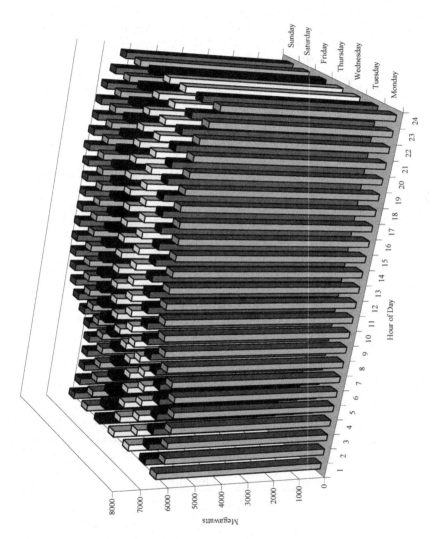

Figure 10.9: Alberta Power, Average Hourly Load

10.5   Empirical Evidence

A matter that has to be dealt with before we could proceed to perform Granger causality tests concerns the lengths of lags r, s, and q in equation (10.1). In the literature r, s, and q are frequently chosen to have the same value and lag lengths of 4, 6, or 8 are used, for example, most often with quarterly data. Given the hourly frequency of our data, we report p-values for Granger causality tests in Tables 10.3 (for all hours), 10.4 (for peak hours), and 10.5 (for off peak hours) for five different lag lengths — 50, 100, 150, 200, and 250. p-values less than .001 indicate rejection of the null hypothesis of no causality at the 1% level.

The results with the following test statistics are provided in Tables 10.3-10.5. The statistic η_1 is the asymptotic F-test statistics for the null hypothesis that Imports do not cause power prices, when the coefficients of Exports are not restricted to equal zero. The statistic η_2 is the asymptotic F-test statistics for the null hypothesis that Exports do not cause power prices, when the coefficients of Imports are not restricted to equal zero. The statistic η_3 is the asymptotic F-test statistics for the null hypothesis that Imports and Exports jointly do not cause power prices.

TABLE 10.3

MARGINAL SIGNIFICANCE LEVELS OF GRANGER
CAUSALITY TESTS: ALL HOURS

$$\text{Price}_t = a_0 + a_1 t + \sum_{j=1}^{r} \alpha_j \text{Price}_{t-j} + \sum_{j=1}^{s} \beta_j \text{Imports}_{t-j} +$$

$$+ \sum_{j=1}^{q} \gamma_j \text{Exports}_{t-j} + \sum_{j=1}^{p} \delta_j \text{Load}_{t-j} + \varepsilon_t$$

		Test statistics	
Lag	η_1 ($\beta_j = 0$ for all j)	η_2 ($\gamma_j = 0$ for all j)	η_3 ($\beta_j = \gamma_j = 0$ for all j)
50	<.001	0.075	<.001
100	<.001	<.001	<.001
150	<.001	<.001	<.001
200	0.338	<.001	<.001
250	<.001	<.001	<.001

TABLE 10.4

MARGINAL SIGNIFICANCE LEVELS OF GRANGER
CAUSALITY TESTS: PEAK HOURS

$$\text{Price}_t = a_0 + a_1 t + \sum_{j=1}^{r} \alpha_j \text{Price}_{t-j} + \sum_{j=1}^{s} \beta_j \text{Imports}_{t-j} +$$

$$+ \sum_{j=1}^{q} \gamma_j \text{Exports}_{t-j} + \sum_{j=1}^{p} \delta_j \text{Load}_{t-j} + \varepsilon_t$$

| | Test statistics | | |
| | η_1 | η_2 | η_3 |
Lag	($\beta_j = 0$ for all j)	($\gamma_j = 0$ for all j)	($\beta_j = \gamma_j = 0$ for all j)
50	<.001	<.001	<.001
100	0.022	0.016	0.014
150	<.001	<.001	<.001
200	<.001	0.506	<.001
250	<.001	0.045	<.001

TABLE 10.5

MARGINAL SIGNIFICANCE LEVELS OF GRANGER
CAUSALITY TESTS: OFF PEAK HOURS

$$\text{Price}_t = a_0 + a_1 t + \sum_{j=1}^{r} \alpha_j \text{Price}_{t-j} + \sum_{j=1}^{s} \beta_j \text{Imports}_{t-j} +$$

$$+ \sum_{j=1}^{q} \gamma_j \text{Exports}_{t-j} + \sum_{j=1}^{p} \delta_j \text{Load}_{t-j} + \varepsilon_t$$

| | Test statistics | | |
| | η_1 | η_2 | η_3 |
Lag	($\beta_j = 0$ for all j)	($\gamma_j = 0$ for all j)	($\beta_j = \gamma_j = 0$ for all j)
50	<.001	<.001	<.001
100	0.403	<.001	<.001
150	<.001	<.001	<.001
200	<.001	<.001	<.001
250	0.094	<.001	<.001

It appears that the null hypotheses of no causality are in general rejected for all three data sets (that is, all hours, peak hours, and off peak hours). Hence, we arrive at the stylized fact that with our data (that is, hourly prices, load, imports, and exports from January 1, 2000 to July 31, 2005) there is causality from imports and exports to power prices.

10.6   Conclusions

This paper provides a study of the relationship between electricity prices and imports and exports for peak hours, off-peak hours, and all hours, using data over the period from January 1, 2000 to July 31, 2005 from Alberta's (deregulated) spot power market. We find that there are causal relationships from imports and exports to power prices.

In this paper we tested for Granger causality from power imports and power exports to power prices, using a linear model. A logical next step would, therefore, be to investigate causal relationships between these variables in the context of nonlinear models. This is an area for potential productive future research and we are currently investigating such causal relationships using the nonlinear causality test of Baek and Brock (1992) and Hiemstra and Jones (1994), as recently modified by Diks and Panchenko (2005a,b).

Chapter 11

Cointegration Analysis of Power Prices in the Western North American Markets

Mattia Bianchi and Apostolos Serletis

11.1 Introduction

This chapter empirically investigates the extent of integration among the Alberta, Mid Columbia, California Oregon Border, California NP15, and California SP15 power markets. These markets are members of the Western Electricity Coordinating Council (WECC), which is the largest and most diverse of the ten reliability councils that form the North American Electricity Reliability Council (NERC). The WECC is responsible for coordinating and promoting electric system reliability and for facilitating the formation of Regional Transmission Organizations in various parts of the West. Recently, Serletis and Dormaar (2006) have found that inter-tie trade of power among Alberta and British Columbia heavily influences the Alberta's pool price and market dynamics. Price in one market is jointly determined by local market conditions and the conditions of the markets with which it is integrated. Trade is the essential "instrument" that determines market integration; the development of power exchange between different independent markets ideally leads to their integration into a single entity. Hence, the removal of barriers to commercial exchange positively affects the level of integration.

In an ideal economic context, two markets are integrated when a single price exists for the product that is traded on the market (also referred to as the law of one price). However, market imperfections may introduce differences in prices that exceed the arbitrage costs (transportation and transaction costs) that would be included in the law of one price. Thus while testing for market integration, we are not testing for equality among different power price series, but for a long-run equilibrium relationship that links these series together. In the long run prices do not drift apart without limit but a certain relation exists between their time paths. Also, their dynamics in the short run reflect any divergence from the long run relationship.

In the context of power markets, deregulation and open access policies are deemed to lead to market integration. In Alberta, the process of deregulation began in 1996 while the California electric industry was liberalized after April 1998. The open access policies, such as the Energy Policy Act of 1992 and FERC's Orders 888 and 889, obliged transmission-owning utilities to open their transmission systems to market participants, not only generators but also power marketers.

This chapter aims to determine the extent of market integration in the main Western North American power markets and to what extent deregulation and open access to transmission policies have removed barriers to trade among them. In doing so, we test for cointegration between power prices; we develop an error correction model; and then look for causal relationships between the price dynamics in the Alberta, Mid Columbia, California Oregon Border, California NP15, and California SP15 power markets. Evidence of no integration among these interconnected markets would be suggestive of severe transmission constraints or market imperfections.

The chapter is organized as follows. Section 11.2 provides a literature review on the subject of market integration. In Section 11.3 we present the data used in the analysis. Sections 11.4 and 11.5 present the results of the unit root tests and the cointegration tests, respectively. In Section 11.6 we apply the error correction model to the time series and we test for Granger causality. Finally, in Section 11.7 we report the conclusions of the chapter.

11.2   Literature Review

Several research works have analyzed the extent of market integration among natural gas markets — see, for example, De Vany and Walls (1993), King and Cuc (1996), and Serletis (1997), among others. A general finding is that the process of deregulation that begun in the mid 1980's has led to market integration.

There are, however, very few studies of power markets integration. Mc-Collough (1996) finds high price correlations between power markets in WSCC, but not with the Alberta power market and the B.C.-U.S. border. However, the results in this study are misleading, since price correlation is not the appropriate technique to delineate markets. Woo, Lloyd-Zanetti, and Horowitz (1997) test for market integration and for price competition in the Pacific Northwest region of WECC. They use daily peak prices for 1996 from the Mid-Columbia (Mid-C), California Oregon Border (COB), B.C.-U.S. Border, and Alberta markets and report that all four price series are cointegrated.

The results, however, reported by Woo et al. (1997) are spurious, since their price series are found to be stationary and testing for cointegration among stationary series makes no sense, since cointegration is a property of integrated variables. Moreover, the relevance of the Woo et al. (1997) study is low due to data limitations (their sample size consists of only 252 observations). Finally, as already noted, testing for market integration implies testing for a long-run equilibrium relationship. Since the data are relative to 1996 only, the time horizon is too short to be considered representative of a long-run relationship.

Finally, Bailey (1998) finds that in most of the Western United States the wholesale electricity markets are integrated and thus they constitute a wider market. However, the supply and demand conditions which create congestion along transmission lines, such as transmission line outages and de-ratings, and high demand and high hydroelectric flows from the Pacific Northwest, cause the expanse of the geographic market to narrow at certain times, since transmission congestion prevents economical trade from occurring. Hence, she concludes that the geographic expanse of the power market in Western North America is dynamic and changes in response to shocks to supply and demand. For the majority of the time (80%) from June 1995 to December 1996, arbitrage constraints bind prices in the Pacific Northwest and California electricity markets, while transmission congestion causes prices separation in 19% of the observations and autarky prevails in the remaining 1%.

De Vany and Walls (1999a,b) have modelled the dynamic behaviour of prices in a network of interconnected electric power markets. They found market integration and price convergence in five U.S. electricity spot markets. Although their research strongly contributes to the knowledge about this topic, it focuses only on decentralized bilateral markets (very different from the centralized power pools) and uses data from 1994 to 1996. Their results show evidence of an efficient and stable wholesale power market in the western area of United States.

Market conditions in the electricity industry have changed significantly since the time that the research works mentioned above were performed. In particular, restructuring of the Alberta and California power markets has potentially modified the market and price dynamics in the Pacific Northwest area. It is our objective in this chapter to assess whether the findings of the earlier literature (reviewed above) are valid and to provide new evidence, using recent state-of-the-art advances in the field of applied econometrics, regarding market and power price dynamics in the Pacific Northwest.

11.3   The Data

The data used in this chapter consist of daily peak and off-peak power prices for the Alberta, Mid Columbia, California Oregon Border, California North Path 15, and California South Path 15 markets. The sample period is from April 1st 1998 to October 31st 2005. Prices in the three centralized power exchanges (run by an Independent System Operator) — Alberta, California NP15, and California SP15 — are hourly. With regard to the Alberta's power market, we used the hourly real time pool price posted by the Alberta Electric System Operator (AESO). This price is the average of sixty minute-by-minute system marginal prices that are based on the price of the highest bid that must be dispatched to meet pool demand in Alberta. Alberta's pool prices are converted from Canadian dollars to U.S. dollars in order to compare the values with those of the other U.S. markets.

Regarding the California NP15 and SP15 power markets, for the period from April 1st 1998 to September 30th 2004 we used the ISO Hourly Ex Post Price; according to the California ISO (CAISO), the Hourly Ex Post Prices are the "hourly market clearing price." They are the zone-specific averages of the six 10-minute market clearing prices in an hour, weighted by the amount of Instructed Imbalance Energy during each 10-minute interval. For the period from October 1st 2004 to October 31st 2005, we use the ISO Hourly Average Energy Prices, which now represent the market clearing prices, since CAISO overhauled the Real Time Market and dispatching system in October 2004. It is to be noticed that the ISO Hourly Average Energy Prices are energy weighted averages of the zonal market clearing prices over all twelve 5-minute intervals for each zone.

On the other hand, the Mid Columbia and California Oregon Border markets are not centralized power exchanges, but they are trading hubs where power is bilaterally traded among utilities and marketers. Hence, in these markets power prices refer to volume weighted averages of actual transaction prices obtained through daily surveys. Thus the data consist of daily spot market prices for pre-scheduled energy transactions spanning the period from April 1st 1998 to October 31st 2005. Pre-scheduled transactions are energy transactions that are scheduled a day-ahead of actual physical production. These data, which is provided by Platts, consist of two daily observations: peak and off-peak prices. Since Mid-C and COB only have daily peak and off-peak prices, we calculated the daily peak and off-peak prices for Alberta, NP15, and SP15 as equally-weighted averages of the peak and off-peak hourly prices.

For Alberta, we adopt the official WECC definition for peak and off-peak periods. The official WECC definition for peak is the hour ending (HE) 8:00 to the HE 23:00 Monday through Saturday inclusive. The official definition of the off-peak is the remaining hours Monday through Saturday, Sundays, and statutory holidays. For California NP15 and SP15, we adopt the CAISO definition; that is, from HE 7:00 to the HE 22:00. The definition of peak hours in Mid-Columbia and California Oregon Border is from HE 7:00 to HE 22:00, Monday through Saturday, prevailing Pacific Time. Note that by taking into account the time difference, the definition of peak hours in Alberta (Mountain Time) and the U.S. Pacific markets (Pacific Time) perfectly matches. Thus Alberta, NP15, and SP15 peak price is defined as the average of the hourly prices during the peak period of the day (i.e. 16 hours); the corresponding off-peak price is defined as the average of the hourly prices during the off-peak period of the day (i.e. 8 hours from Mondays to Saturdays, 24 hours on Sundays).

Thus, we obtain two data sets. Peak prices for Alberta, Mid-C, COB, NP15, and SP15 constitute the peak data set, consisting of 2,375 price observations. Off-peak prices for Alberta, Mid-C, COB, NP15, and SP15 constitute the off-peak data set, consisting of 2,491 price observations. Due to missing values in the Mid-C and COB price series, off-peak prices on Sundays from April 1st 1998 to December 31st 2001 are excluded, while prices on Sundays from 2002 to 2005 are included in the data set.

Figures 11.1 and 11.2 show the daily peak and off-peak prices in the Alberta, Mid-C, COB, NP15, and SP15 power markets. As can be seen in Figure 11.1, wholesale power peak prices move more or less together from April 1998 until the beginning of 2001. During 2001, the price series seem to drift apart: prices in California NP15 and SP15 decrease due to the imposition of low price caps (US$250 first, US$150 later). Alberta prices decrease as well, while Mid-C and COB prices remain very high, mainly due to low precipitation and water supplies and the absence of price caps. From the second half of 2001, prices return to move together, rather constantly around the mean. Alberta's market shows the most frequent price spikes, due to a higher price cap (Can$1,000) than California's one.

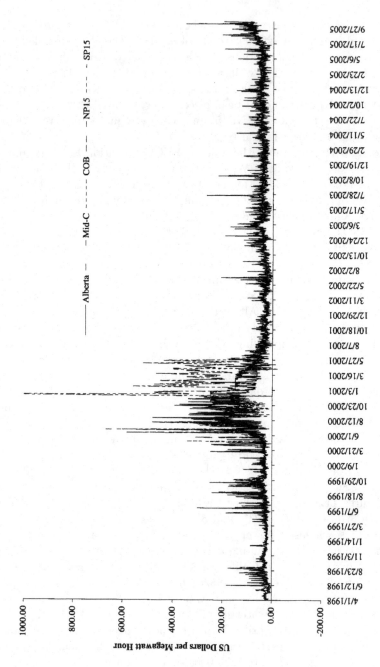

Figure 11.1: Peak Prices

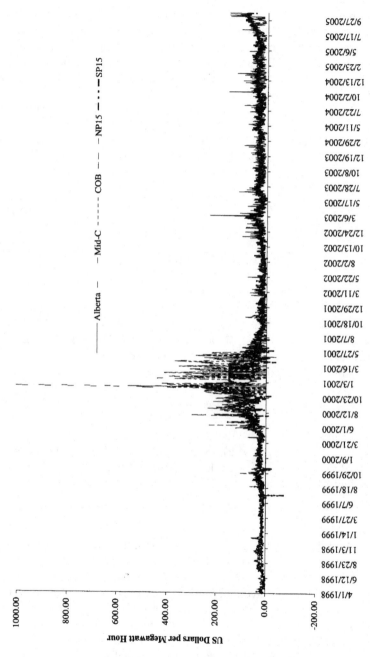

Figure 11.2: Off-Peak Prices

With regard to off-peak prices, Alberta has significantly the lowest prices, also due to the different off-peak definition that better reflects the pattern of power consumption. SP15 has the lowest price among the U.S. markets. The Mid-C bilateral market, however, seems to be performing very badly, with the highest average price, standard deviation, and coefficient of variation. Figure 11.2, basically, shows similar patterns to those of Figure 11.1. Again, prices move more or less together from April 1998 until the end of 2000, when NP15 presents the highest values. Then, at the beginning of 2001, prices in California are capped by regulators, while prices in Mid-C and COB spike up to $1,000 until October 2001. Since then, prices in all markets have decreased and so has the volatility.

TABLE 11.1
DATA SUMMARY

	Alberta	Mid-C	COB	NP15	SP15
A. Peak Prices					
Observations	2375	2375	2375	2375	2375
Mean	56.06	64.36	66.31	54.38	53.17
Standard deviation	52.42	119.65	101.38	54.92	50.72
Coefficient of variation	0.94	1.86	1.53	1.01	0.95
B. Off-peak Prices					
Observations	2491	2491	2491	2491	2491
Mean	27.18	47.02	41.05	41.71	34.86
Standard deviation	21.06	71.22	49.01	42.45	33.38
Coefficient of variation	0.77	1.51	1.19	1.02	0.96

Panel A of Table 11.1 reports the mean, standard deviation, and coefficient of variation of the peak price series, while Panel B reports the same statistics for off-peak prices. On average, the peak price for one megawatt-hour of electricity in the five markets over the period from April 1st 1998 to October 31st 2005 is around US$60. Power in Southern California is the cheapest while power peak prices at COB are the highest. These results are quite surprising: not only the ISO centralized markets of Alberta and California, which are deemed to have more frequent price spikes, have the lowest power prices but also the lowest volatility. The reason for that is the existence of price caps in the California and Alberta centralized markets. Being decentralized bilateral markets, COB and Mid-C do not have price caps, thus allowing price excursions up to $4,000. The coefficient of variation for Alberta and California is around one, while the same coefficient for Mid-C and COB is much higher, 1.86 and 1.53, respectively.

11.4 Testing for Stochastic Trends

We start by testing for the presence of a stochastic trend (a unit root) in the autoregressive representation of each individual time series. A time series that has a stochastic trend is said to be non-stationary. Most economic and financial variables that exhibit strong trends, like GDP and price levels, are non-stationary and thus have a unit root. In many cases, the first difference of a non-stationary time series is stationary. When this is true, the time series is said to be integrated of order one [or I(1) in the terminology of Engle and Granger (1987)]. More generally, a non-stationary time series is integrated of order n, or I(n), if it turns out to be stationary after being differenced n times. Conversely, a stationary time series is integrated of order zero, or I(0).

In order to test for the existence of a stochastic trend, we use the augmented Dickey-Fuller (ADF) test — see Dickey and Fuller (1981). Thus, we test the null hypothesis of a stochastic trend by estimating the following ADF regression equation

$$\Delta x_t = \alpha_0 + \alpha_1 t + \gamma x_{t-1} + \sum_{j=1}^{k} \beta \Delta x_{t-j} + \varepsilon_j$$

where x represents a price variable, Δ is the difference operator, and k is the optimal lag length, determined using the AIC+2 rule suggested by Pantula, Gonzalez-Farias, and Fuller (1994).

Tables 11.2 and 11.3 present the results of the ADF unit root tests for the peak and off-peak prices, respectively. In panel A of each table the test is applied to the levels of the series and in panel B to the differences of the series. Each row corresponds to one market. The second column of each table reports the optimal lag length, the third the t-statistic for the null hypothesis $\gamma = 0$, and fourth column shows the p-value for the null hypothesis of a unit root. Finally, the fifth column summarizes the outcome of the test for each market, that is whether the price series has a unit root or not.

TABLE 11.2

ADF UNIT ROOT TEST RESULTS FOR PEAK HOUR PRICES

Series	Number of lags	t-statistic	p-value	Decision
A. Levels of the Series				
Alberta	27	-2.663	0.252	I(1)
Mid-C	14	-5.473	<0.001	I(0)
COB	14	-4.749	<0.001	I(0)
NP15	27	-2.335	0.415	I(1)
SP15	27	-3.011	0.129	I(1)
B. Differences of the Series				
Alberta	27	-14.448	<0.001	I(0)
Mid-C				
COB				
NP15	27	-15.183	<0.001	I(0)
SP15	27	-14.409	<0.001	I(0)

TABLE 11.3

ADF UNIT ROOT TEST RESULTS FOR OFF-PEAK HOUR PRICES

Series	Number of lags	t-statistic	p-value	Decision
A. Levels of the Series				
Alberta	26	-3.302	0.066	I(1)
Mid-C	17	-4.083	0.007	I(0)
COB	27	-2.726	0.225	I(1)
NP15	19	-2.737	0.221	I(1)
SP15	27	-3.373	0.055	I(1)
B. Differences of the Series				
Alberta	27	-11.505	<0.001	I(0)
Mid-C				
COB	27	-13.888	<0.001	I(0)
NP15	18	-14.806	<0.001	I(0)
SP15	27	-11.338	<0.001	I(0)

First, we look at the results for the levels of the series. At the 5% significance level, the null hypothesis of a unit root is rejected if the p-value is less than 0.05. The null hypothesis cannot be rejected for the Alberta and California NP15 and SP15 peak price series. We conclude that these variables have a stochastic trend. Conversely, the null hypothesis is rejected for the Mid-C and COB peak prices series. These variables appear to be stationary. These results are interesting: basically, the price series of the centralized power markets have a stochastic trend, while the price series of Mid-C and COB are stationary. We attribute this difference in results to the different price formation processes: in Alberta and California, the spot price is formed real time by the market forces and the pool price is not known until after the fact. On the other hand, Mid-C and COB are bilateral power markets where prices are known prior to delivery of the energy and most trades are transacted day ahead over on- and off-peak strips.

In our view, different price formation processes cause the prices to have a deterministic or stochastic behavior in these markets. While in Alberta and California, deviations of power prices from their underlying trend appear to be permanent, in Mid-C and COB, the fluctuations in power prices are viewed as temporary since prices are expected to return to their trend growth rate in the long run. Subsequently, in order to determine the order of integration for the Alberta, NP15, and SP15 power price series, which turned out to be non-stationary, we apply the ADF test to the first differences of the series. Since all the p-values are less than 0.05, we reject the null hypothesis and we conclude that the differenced series are stationary.

With regard to off-peak price series, the null hypothesis of a unit root is rejected only for the Mid-C off-peak prices. Alberta, COB, NP15, and SP15 off-peak prices appear to have a unit root and, after running the same test to the differenced series, we conclude that these series are integrated of order 1, or I(1).

11.5 Testing for Cointegration

Since cointegration is a property of non-stationary series, in this section we test for cointegration in the Alberta, NP15, and SP15 markets, using peak prices as well as off-peak prices. The concept of cointegration, first introduced by Engle and Granger (1987), refers to a linear combination of non-stationary variables that is itself stationary. In particular, two non-stationary series x and y are said to be cointegrated if there exists a linear combination

$$\varepsilon_t = y_t - \alpha - \beta x_t$$

that is stationary. We test for cointegration between Alberta, NP15, and SP15 power prices using the Engle and Granger (1987) method. That is, we test for a unit root in the regression residuals, using the ADF test and appropriate critical values in order to take into account the number of variables in the regression — see, for example, Serletis and Herbert (1999).

If the null hypothesis of no cointegration (or equivalently of stationarity of $\hat{\varepsilon}_t$) is rejected, the price series cointegrate and thus there is a long run relationship between these series. Since cointegration test results can be sensitive to the roles of each market as dependent and independent variables, we reverse the roles in the regression. In addition to that, we run six trivariate regressions in which each price series is treated as the dependent variable and the remaining two as independent variables, in the context of the following model

$$x_{it} = \alpha + \beta x_{jt} + \gamma x_{kt} + \varepsilon_t$$

TABLE 11.4

ENGLE-GRANGER BIVARIATE COINTEGRATION TEST
RESULTS FOR PEAK HOUR PRICES

Model: $x_{it} = \alpha + \beta x_{jt} + \varepsilon_t$					
Series pair	Dependent Variable	Number of lags	t-statistic	p-value	Decision
Alberta, NP15	NP15	27	-4.226	0.003	Cointegration
	Alberta	19	-7.034	<0.001	Cointegration
Alberta, SP15	SP15	27	-4.702	<0.001	Cointegration
	Alberta	22	-5.387	<0.001	Cointegration
NP15, SP15	SP15	27	-5.477	<0.001	Cointegration
	NP15	27	-4.353	0.002	Cointegration

Table 11.4 presents the results for the Engle-Granger bivariate cointegration tests applied to peak hour prices. Each series is tested against the other series, both as dependent variable and as independent variable. Clearly, we get no contradiction: the null hypothesis of no cointegration is rejected

at the 1% level for all pairs of series. Thus, peak hour power prices in Alberta and California NP15 and SP15 are cointegrated, meaning that they share stochastic trends and that a long-run equilibrium relationship exists among these prices. Table 11.5 confirms these results using trivariate regression tests; that is, we find strong evidence of cointegration among Alberta, NP15, and SP15 peak hour power prices. Using off-peak prices, we get essentially the same results (see Tables 11.6 and 11.7) and thus we conclude that there is strong evidence of market integration among Alberta, NP15, and SP15 power markets — that is, we interpret rejections of the null hypothesis of no power price cointegration as evidence of market integration.

One first interpretation of the results is that these three power markets are linked together, since the same underlying stochastic growth components are apparently affecting their price dynamics. Moreover, transmission capacity does not appear to prevent power trade from occurring between these markets. Not only there is a high degree of market integration during off-peak periods, when congestion is unlikely to occur, but also during peak periods. Also, these results show that unexploited profit opportunities from trade among these power markets are not likely to exist. Arbitrage seems to work even though the distances between the regions are large and traverse several utility service areas, where they normally incur transmission tariffs. In other words, since the integrated price series cointegrate, the price differentials are stationary. Thus, there is price convergence and arbitrage mechanisms and price competition disciplines prices. Every permanent shock in the trend of one price series is ultimately transmitted to the trend of the other price series.

TABLE 11.5

ENGLE-GRANGER TRIVARIATE COINTEGRATION TEST
RESULTS FOR PEAK HOUR PRICES

Model: $x_{it} = \alpha + \beta x_{jt} + \gamma x_{kt} + \varepsilon_t$				
Dependent Variable	Number of lags	t-statistic	p-value	Decision
Alberta	19	-7.149	<0.001	Cointegration
NP15	27	-4.945	<0.001	Cointegration
SP15	27	-5.535	<0.001	Cointegration

TABLE 11.6

ENGLE-GRANGER BIVARIATE COINTEGRATION TEST
RESULTS FOR OFF-PEAK HOUR PRICES

Model: $x_{it} = \alpha + \beta x_{jt} + \varepsilon_t$

Series pair	Dependent Variable	Number of lags	t-statistic	p-value	Decision
Alberta, NP15	NP15	24	-4.554	0.001	Cointegration
	Alberta	24	-6.99	<0.001	Cointegration
Alberta, SP15	SP15	27	-5.590	<0.001	Cointegration
	Alberta	25	-6.607	<0.001	Cointegration
NP15, SP15	SP15	27	-6.146	<0.001	Cointegration
	NP15	28	-4.746	<0.001	Cointegration

TABLE 11.7

ENGLE-GRANGER TRIVARIATE COINTEGRATION TEST
RESULTS FOR OFF-PEAK HOUR PRICES

Model: $x_{it} = \alpha + \beta x_{jt} + \gamma x_{kt} + \varepsilon_t$

Dependent Variable	Number of lags	t-statistic	p-value	Decision
Alberta	24	-7.128	<0.001	Cointegration
NP15	28	-4.779	0.002	Cointegration
SP15	27	-6.279	<0.001	Cointegration

Given the results of the cointegration analysis, we can proceed to test for strong market integration. Strong market integration implies that price changes in separate markets track one another with unitary responses, so that prices contain and reflect the same information. In the case of strong market integration, price shocks at one region are proportionally reflected in all other market prices. Note that price levels may differ across regions reflecting the shadow value of transmission capacity, though relative

changes between prices across market pairs being equal. Now for the price
pairs that cointegrate, we proceed on estimating β in the cointegrating regression, which represents the factor of proportionality between the shared
stochastic trends. Estimates of β significantly equal to 1 would suggest
evidence of strong market integration.

TABLE 11.8

ESTIMATES OF FACTORS OF PROPORTIONALITY β
FOR SHARED STOCHASTIC TRENDS

| | *Dependent Variable* | | | | | |
| | Alberta | | NP15 | | SP15 | |
	β	$t(\beta)$	β	$t(\beta)$	β	$t(\beta)$
A. Peak hours						
Alberta			0.647	4.226	0.535	4.702
NP15	0.587	7.034			0.864	5.477
SP15	0.571	5.387	1.013	4.353		
B. Off-peak hours						
Alberta			1.169	4.554	0.812	5.590
NP15	0.304	6.999			0.663	6.146
SPS15	0.347	6.607	1.090	4.746		

As can be seen in Table 11.8, the estimates of β for the California
NP15 and SP15 markets during peak and off-peak periods are close to
unity (1.013 and 1.090, respectively), suggesting the existence of strong
market integration — since estimates of β depend on the choice of the
dependent variable, for each market pair we discuss the β with the higher
value. However, market integration between the California markets and
Alberta is weaker. During peak periods (see panel A of Table 11.8), the
factors of proportionality are estimated to be equal to 0.647 and 0.571 for
NP15 and SP15, respectively. Such results are reasonable given the longer
distance, the different market structures and rules, and the existence of
other power markets between Alberta and California. However, during off-
peak hours (see panel B of Table 11.8), the extent of market integration
increases; that is, the estimates of β are closer to unity (1.169 and 0.812
for Alberta-NP15 and for Alberta-SP15, respectively). Also, in both peak
and off-peak periods, Alberta shows a stronger integration with NP15 than

with SP15. This makes sense, due to the shorter distance from NP15 and to the stronger influence that power prices in the U.S. southwest markets (i.e. Palo Verde and Four Corners) have on SP15 power prices.

In addition to the factors of proportionality, β, the constants, α, can also be estimated for the market-pairs that we have established cointegration. α represents the average transmission and transactions costs and the line losses. If two markets are integrated, the difference in mean prices would reflect, at most, the average transmission and transaction costs — see, for example, Woo *et al.* (1997). Two markets are perfectly integrated if the corresponding prices are perfect predictors of one another in expectation. Formally, there is perfect integration if in the cointegrating relation $\beta = 1$ (condition for strong market integration) and $\alpha = 0$. Table 11.9 reports the estimates of α, together with the t-statistics, for Alberta, NP15, and SP15 peak (see panel A) and off-peak (see panel B) power prices — again, since the estimates of α depend on the choice of the dependent variable, for each market pair we discuss the α with the lower value.

TABLE 11.9

ESTIMATES OF CONSTANT α

IN THE COINTEGRATING REGRESSION

	Dependent Variable					
	Alberta		NP15		SP15	
	β	$t(\beta)$	β	$t(\beta)$	β	$t(\beta)$
			A. Peak hours			
Alberta			18.121	13.956	23.182	18.249
NP15	24.011	20.158			6.190	11.937
SP15	25.688	19.776	0.544	0.941		
			B. Off-peak hours			
Alberta			10.135	9.569	12.776	14.656
NP15	14.652	30.974			7.081	14.915
SPS15	15.316	29.734	4.022	6.403		

First, we do not find evidence of perfect market integration between the California NP15 and SP15 markets, since the estimates of the arbitrage costs are not equal to zero. During peak hours, the transmission and transaction costs are about \$0.54, while during off-peak hours these costs increase quite surprisingly. That might be explained by the fact there is less substitution across markets during peak periods and thus more power trades and flows occur during off-peak periods. Since the cost of transmission is proportional to the amount of energy flowing on the transmission line, the estimates of α are higher during off-peak hours. On the other hand, the arbitrage costs between Alberta and California are fairly high, ranging from around \$10 to \$23. They reflect greater line losses, which are proportional to the distance. In fact, α is higher in the Alberta-SP15 regression than in the Alberta-NP15 regression, since southern California is further from Alberta than northern California. Also, the transaction costs are higher since the tie-lines that connect the Alberta and California regions are owned and managed by several utilities. Generally, we find that the average transmission costs between Alberta and California are higher during peak periods.

11.6 Error Correction Modeling and Causality Testing

Since the power price series in the Alberta and California NP15 and SP15 markets are cointegrated, a long run equilibrium relationship exists between these series. The time paths of these cointegrated variables in the short run are affected by any deviation from the long-run equilibrium. In order to return to the long run equilibrium, the movements of the variables in the short run depend on the extent and the direction of the divergence. For instance, if the gap between two time series is small relative to the long run relationship, adjustments of one or both of the variables will re-establish the equilibrium by widening the gap. According to the Engle and Granger representation theorem, the short run dynamics can be described by an error correction model, relating current and lagged first differences of y_t and x_t and at least one lagged value of $\hat{\varepsilon}_t$.

11.6.1 Bivariate Granger Causality Tests

According to Engle and Granger (1987), the error-correction model is of the following form

$$\Delta y_t = \alpha_1 + \alpha_y \hat{\varepsilon}_{t-1} + \sum_{j=1}^{r} \alpha_{11}(j)\Delta y_{t-j} + \sum_{j=1}^{s} \alpha_{12}(j)\Delta x_{t-j} + \varepsilon_{yt} \quad (11.1)$$

$$\Delta x_t = \alpha_2 + \alpha_x \hat{\varepsilon}_{t-1} + \sum_{j=1}^{r} \alpha_{21}(j)\Delta y_{t-j} + \sum_{j=1}^{s} \alpha_{22}(j)\Delta x_{t-j} + \varepsilon_{xt} \quad (11.2)$$

where α_1, $\alpha_2, \alpha_y, \alpha_x, \alpha_{11}(j), \alpha_{12}(j), \alpha_{21}(j)$, and $\alpha_{22}(j)$ are all parameters, ε_{yt} and ε_{xt} are white noise disturbances and $\hat{\varepsilon}_{t-1}$ is the error correction term and estimates the deviation from long run equilibrium in period $t-1$. The error correction model focuses on the short run dynamics while making them consistent with the long run equilibrium. It shows how y_t and x_t change in response to stochastic shocks (represented by ε_{yt} and ε_{xt}) and to the previous period's deviation from the long-run equilibrium (represented by $\hat{\varepsilon}_{t-1}$).

To test for Granger causality from x_t to y_t, we first fit equation (11.1) by OLS to obtain the unrestricted sum of squared residuals, SSR_u. Then we run another regression equation under the null hypothesis that α_y and all the coefficients of the lagged values of Δx_t are zero, to obtain the restricted sum of squared residuals, SSR_r. The statistic

$$\frac{(SSR_r - SSR_u)/(s+1)}{SSR_u/(T-r-s-2)}$$

has an asymptotic F-distribution with numerator degrees of freedom $(s+1)$ and denominator degrees of freedom $(T-r-s-2)$. T represents the number of observations, s is the number of lags for Δx_t in equation (11.1), and the number 2 is subtracted in order to take in account for the constant term and the error correction term. If the null hypothesis cannot be rejected, then we conclude that the data do not show causality. If the null hypothesis is rejected, then we conclude that the data do show causality. The same procedure is reversed in another F-test to assess whether a feedback relationship exists between these series, as in the context of equation (11.2). In determining the optimal values of r and s in each of equations (11.1) and (11.2), we allow a maximum value of 60 for each r and s and by running 3,600 regressions for each bivariate relationship we choose the lag length that produces the smallest value of the AIC.

Panels A and B of Table 11.10 report the results of the bivariate Granger causality tests applied to peak and off-peak series, respectively. The null hypothesis of no causality is rejected in all cases. Hence, for all market pairs, we find evidence of significant bidirectional causality during peak and off-peak periods. In other words, knowledge of past (say) Alberta power prices improves the prediction of future (say) NP15 power prices beyond predictions that are based on past NP15 power prices alone.

TABLE 11.10

MARGINAL SIGNIFICANCE LEVELS FOR GRANGER CAUSALITY TESTS
IN THE CONTEXT OF BIVARIATE ERROR CORRECTION MODELS

$$\Delta y_t = \alpha_1 + \alpha_y \hat{\varepsilon}_{t-1} + \sum_{j=1}^{r} \alpha_{11}(j)\Delta y_{t-j} + \sum_{j=1}^{s} \alpha_{12}(j)\Delta x_{t-j} + \varepsilon_{yt} \qquad (11.1)$$

$$\Delta x_t = \alpha_2 + \alpha_x \hat{\varepsilon}_{t-1} + \sum_{j=1}^{r} \alpha_{21}(j)\Delta y_{t-j} + \sum_{j=1}^{s} \alpha_{22}(j)\Delta x_{t-j} + \varepsilon_{xt} \qquad (11.2)$$

	Alberta			NP15			SP15		
Series	Optimal lag	F-statistic	p-value	Optimal lag	F-statistic	p-value	Optimal lag	F-statistic	p-value
A. Peak hours									
Alberta									
NP15	(54,59)	4.036	<0.001	(35,41)	2.471	<0.001	(60,57)	2.753	<0.001
SP15	(57,60)	4.480	<0.001	(55,55)	6.715	<0.001	(53,58)	7.660	<0.001
B. Off-peak hours									
Alberta									
NP15	(55,16)	2.170	0.005	(37,47)	2.327	<0.001	(56,60)	2.278	<0.001
SPS15	(55,16)	2.189	0.004	(55,57)	3.292	<0.001	(55,60)	3.923	<0.001

11.6.2   Trivariate Granger Causality Tests

We also tested for Granger causality, in the context of the following trivariate models

$$\Delta y_t = \alpha_1 + \alpha_y \hat{\varepsilon}_{t-1} + \sum_{j=1}^{r} \alpha_{11}(j) \Delta y_{t-j}$$

$$+ \sum_{j=1}^{s} \alpha_{12}(j) \Delta x_{t-j} + \sum_{j=1}^{q} \alpha_{13}(j) \Delta z_{t-j} + \varepsilon_{yt} \qquad (11.3)$$

$$\Delta x_t = \alpha_2 + \alpha_x \hat{\varepsilon}_{t-1} + \sum_{j=1}^{r} \alpha_{21}(j) \Delta y_{t-j}$$

$$+ \sum_{j=1}^{s} \alpha_{22}(j) \Delta x_{t-j} + \sum_{j=1}^{q} \alpha_{23}(j) \Delta z_{t-j} + \varepsilon_{xt} \qquad (11.4)$$

$$\Delta z_t = \alpha_3 + \alpha_z \hat{\varepsilon}_{t-1} + \sum_{j=1}^{r} \alpha_{31}(j) \Delta y_{t-j}$$

$$+ \sum_{j=1}^{s} \alpha_{32}(j) \Delta x_{t-j} + \sum_{j=1}^{q} \alpha_{33}(j) \Delta z_{t-j} + \varepsilon_{zt} \qquad (11.5)$$

In equations (11.3)-(11.5), the coefficients are defined as those in equations (11.1)-(11.2) and the optimal lag lengths have been determined as those in equations (11.1) and (11.2). However, due to the more computational power needed, in equations (11.3), (11.4), and (11.5) we allowed a maximum value of 24 for each of r, s, and q, thereby running 13,824 regressions for each bivariate relationship in order to determine the optimal lag length.

The results from the trivariate error correction models, reported in Table 11.11, confirm the evidence of joint bidirectional causality. Essentially, knowledge of past (say) Alberta and SP15 power prices improves the prediction of future (say) NP15 prices beyond predictions that are based on past NP15 prices alone. Note that Granger causality refers merely to predictability and has no implications for the strength of conclusions which refer to underlying structural factors — see, for example, Serletis and Herbert (1999). Moreover, according to Woo et al. (1997), causality tests determine whether price behaviour within the individual submarkets that results in

TABLE 11.11
MARGINAL SIGNIFICANCE LEVEL FOR GRANGER CAUSALITY TESTS IN THE CONTEXT OF TRIVARIATE ERROR CORRECTION MODELS

$$\Delta y_t = \alpha_1 + \alpha_y \hat{\varepsilon}_{t-1} + \sum_{j=1}^{r} \alpha_{11}(j)\Delta y_{t-j} + \sum_{j=1}^{s} \alpha_{12}(j)\Delta x_{t-j} + \sum_{j=1}^{q} \alpha_{13}(j)\Delta z_{t-j} + \varepsilon_{yt} \qquad (11.3)$$

$$\Delta x_t = \alpha_2 + \alpha_x \hat{\varepsilon}_{t-1} + \sum_{j=1}^{r} \alpha_{21}(j)\Delta y_{t-j} + \sum_{j=1}^{s} \alpha_{22}(j)\Delta x_{t-j} + \sum_{j=1}^{q} \alpha_{23}(j)\Delta z_{t-j} + \varepsilon_{xt} \qquad (11.4)$$

$$\Delta z_t = \alpha_3 + \alpha_z \hat{\varepsilon}_{t-1} + \sum_{j=1}^{r} \alpha_{31}(j)\Delta y_{t-j} + \sum_{j=1}^{s} \alpha_{32}(j)\Delta x_{t-j} + \sum_{j=1}^{q} \alpha_{33}(j)\Delta z_{t-j} + \varepsilon_{zt} \qquad (11.5)$$

Dependent Variable	Number of lags (r, s, q)	η_1	p-value	η_2	p-value
A. Peak hours					
Alberta	(20,16,20)	3.367	<0.001	5.362	<0.001
NP15	(1,18,22)	5.283	0.005	5.865	<0.001
SP15	(24,20,22)	6.713	<0.001	11.542	<0.001
B. Off-peak hours					
Alberta	(20,18,22)	3.154	<0.001	2.644	0.001
NP15	(19,20,21)	2.800	<0.001	3.953	<0.001
SPS15	(18,24,13)	3.291	<0.001	3.514	<0.001

harmonious overall price movements is suggestive of price leadership. Since
each pair of integrated markets is linked together by a bidirectional feedback
relationship, the price change in one market instantaneously affects prices
in the other market and vice versa. Thus we conclude that there is no price
leadership, suggesting the existence of price competition in the Western
Electricity Coordinating Council.

11.7   Conclusions

We have found that different market structures heavily affect price formation processes. In particular, by testing for unit roots, we found the
presence of a stochastic trend only in the price series that refer to centralized power pools; that is, Alberta and California NP15 and SP15. In these
markets, power is exchanged real time in an auction mechanism by match-
ing actual supply and demand. Such a pool price is known only after-the
fact. Conversely, prices in the Mid-C and COB decentralized markets refer
to transactions that take place day-ahead of actual physical production.
Also, power is not exchanged through pools and auction mechanisms but
through bilateral contracts between generators, utilities, and marketers. As
a consequence of the different price formation, we found strong evidence of
stationarity in the Mid-C and COB peak power price series. Hence, shocks
to power prices in Alberta and California are permanent while shocks to
power prices in Mid-C and COB appear to be temporary. However, stochastic trend behaviour does not appear to infer high volatility and higher
prices and price spikes, since the Alberta and California deregulated markets enjoy lower prices and volatility than the Mid-C and COB markets
do.

   We have also investigated the extent of market integration in the Pacific electricity markets in the WECC. The outcome of this analysis is that
Alberta and California power markets are significantly integrated, since
a long-run equilibrium relationship exists among their prices. This means
that there are empirically effective arbitrage mechanisms that bind the price
movements across these markets, although transfer capacities are limited
in some parts of the WECC grid. Path 15 and the Alberta-BC interconnection are generally considered as transmission bottlenecks that limit
unconstrained power trade. This fact, together with differences in market
structures and regulatory regimes, reduce the extent of market integration.

   Indeed, the cointegration analysis suggests that market integration is
stronger between the adjacent NP15 and SP15 power markets, which share
the same market structure and regulatory regimes. Higher transaction and
transmission costs due to longer distance and more complex trading agree-

ments characterize the long run equilibrium between Alberta and the California markets. However, the extent of market integration is significant. The estimation of error-correcting causality models for the integrated and cointegrated price series also revealed causality and a feedback relationship between any two market pairs. These findings seem to suggest the absence of price leadership from any of the markets.

In conclusion, according to our analysis, the deregulated western electricity industry appears to perform well with regard to power and transmission pricing. Unexploited arbitrage opportunities and monopoly pricing of transmission do not seem to exist. Competition apparently works; that is, transmission rates rise during peak hours due to higher load and energy requirements. Under normal conditions, wholesale power customers can turn to generators and utilities dispersed over a wide geographic area in order to buy or sell electricity either within their jurisdictions, in directly connected regions, or in more remote control areas. Finally, we can conclude that an aggregate integrated market for wholesale electricity exists in the Western North America, spanning from Alberta to the U.S. Pacific area.

Part 4

Crude Oil, Natural Gas, and Electricity Markets

Overview of Part 4

Apostolos Serletis

The following table contains a brief summary of the contents of the chapters in Part 4 of the book. This part of the book consists of three chapters addressing a number of issues regarding crude oil, natural gas, and electricity markets.

Crude Oil, Natural Gas, and Electricity Markets

Chapter Number	Chapter Title	Contents
12	The Cyclical Behavior of Monthly NYMEX Energy Prices	This chapter investigates the basic stylized facts of crude oil, heating oil, unleaded gasoline, and natural gas price movements, using the methodology suggested by Kydland and Prescott (1990). It shows that energy prices are in general procyclical.
13	The Message in North American Energy Prices	Chapter 13 explores the degree of shared trends in natural gas, fuel oil, and power prices in the mid-Atlantic area of eastern Pennsylvania, New Jersey, Maryland, and Delaware.
14	Testing for Common Features in North American Energy Markets	This chapter uses the testing procedures recently suggested by Engle and Kozicki (1993) and Vahid and Engle (1993) and investigates the strength of shared trends and shared cycles between West Texas Intermediate oil prices and Henry Hub natural gas prices.

Chapter 12:

This chapter systematically investigates the basic stylized facts of energy price movements using monthly data for the period that energy has been traded on organized exchanges and the methodology suggetsed by Kydland and Prescott (1990). The results indicate that energy prices are in general procyclical, in contrast to the accepted fact that energy prices are countercyclical and leading the cycle.

Chapter 13:

How similar is the price behavior of North American natural gas, fuel oil, and power prices? Using current state-of-the-art econometric methodology, this chapter explores the degree of shared trends across North American energy markets. Across these markets, there appear to be effective arbitrage mechanisms for the price of natural gas and fuel oil, but not for the price of electricity.

Chapter 14:

Using recent advances in the field of applied econometrics, this chapter explores the strength of shared trends and shared cycles between North American natural gas and crude oil markets. In doing so, it uses daily data from January 1991 to April 2001 on spot U.S. Henry Hub natural gas and WTI crude oil prices. The results show that there has been 'decoupling' of the prices of these two sources of energy as a result of oil and gas deregulation in the United States. It also investigates the interconnectedness of North American natural gas markets and finds that North American natural gas prices are largely defined by the U.S. Henry Hub price trends.

Chapter 12

The Cyclical Behavior of Monthly NYMEX Energy Prices

*Apostolos Serletis and Todd Kemp**

12.1 Introduction

The cyclical behavior of energy prices is important and has been the subject of a large number of studies, exemplified by Hamilton (1983). These studies have, almost without exception, concentrated on the apparently adverse business-cycle effects of oil price shocks. For example, Hamilton (1983) working on pre-1972 data and based on vector autoregression (VAR) analysis, concluded that energy prices are countercyclical and lead the cycle. However, as Mork (1988, p. 74) put it

> "... his study pertained to a period in which all the large oil price movements were upward, and thus it left unanswered the question whether the correlation persists in periods of price decline."

In fact, as shown by Mork (1988), there is an asymmetry in the responses in that the correlation between oil price decreases and gross national product (GNP) growth is significantly different than the correlation between oil price increases and GNP growth, with the former being perhaps zero.

---

*Originally published in *Energy Economics* 20 (1998), 265-271. Reprinted with permission.

The objective of this chapter is to examine the cyclical behavior of energy prices using monthly data for crude oil, heating oil, unleaded gasoline and natural gas for the period that each of these commodities has been traded on organized exchanges. In doing so, we follow Lucas (1977) and define the growth and cycle components of a variable as its smoothed trend and the deviation of the smoothed trend from the actual values of the variable, respectively. Moreover, we define energy cycle regularities as the dynamic comovements of the cyclical components of energy prices and the cycle. In particular, the type of business cycle regularities that we consider are autocorrelations and dynamic cross-correlations between the cyclical components of energy prices, on the one hand, and the cycle, on the other. The robustness of the results to alternative measures of the cycle is also investigated.

The chapter is organized as follows. Section 12.2 briefly discusses the Hodrick-Prescott (HP) filtering procedure for decomposing time series into long-run and business cycle components. Section 12.3 discusses the data and presents HP empirical correlations of energy prices with U.S. output, prices and the unemployment rate. Section 12.4 summarizes and concludes the chapter.

12.2   Methodology

For a description of the stylized facts, we follow the current practice of detrending the data with the Hodrick-Prescott (HP) filter — see Prescott (1986). For the logarithm of a time series X_t, for $t = 1, 2, \ldots, T$, this procedure defines the trend or growth component, denoted τ_t, for $t = 1, 2, \ldots, T$, as the solution to the following optimization problem

$$\min_{\tau_t} \sum_{t=1}^{T} (X_t - \tau_t)^2 + \lambda \sum_{t=2}^{T-1} [(\tau_{t+1} + \tau_t) - (\tau_t - \tau_{t-1})]^2$$

so that $X_t - \tau_t$ is the HP filtered series. For $\lambda = 0$ the growth component is the series and as $\lambda \to \infty$, the growth component approaches a linear trend. In our computations, we set $\lambda = 14,400$, as it has been suggested for monthly data.

We measure the degree of comovement of a series with the pertinent cyclical variable by the magnitude of the correlation coefficient $\rho(j)$, $j \in \{0, \pm 1, \pm 2, \ldots\}$. The contemporaneous correlation coefficient $\rho(0)$ gives information on the degree of contemporaneous comovement between the series and the pertinent cyclical variable. In particular, if $\rho(0)$ is positive, zero, or negative, we say that the series is procyclical, acyclical, or countercyclical, respectively. In fact, for data samples of our size it has been suggested [see, for example, Fiorito and Kollintzas (1994)] that for $0.5 \leq |\rho(0)| < 1$, $0.2 \leq |\rho(0)| < 0.5$ and $0 \leq |\rho(0)| < 0.2$, we say that the series is strongly contemporaneously correlated, weakly contemporaneously correlated and contemporaneously uncorrelated with the cycle, respectively. Also, $\rho(j)$, $j \in \{\pm 1, \pm 2, \ldots\}$, the cross correlation coefficient, gives information on the phase-shift of the series relative to the cycle. If $|\rho(j)|$ is maximum for a negative, zero or positive j, we say that the cycle of the series is leading by j periods the cycle, is synchronous, or is lagging by j periods the cycle, respectively.

The Hodrick-Prescott filter is almost universally used in the real business cycle research program and extracts a long-run component from the data, rendering stationary series that are integrated up to fourth order. HP filtering, however, has been questioned as a unique method of trend elimination — see, for example, King and Rebelo (1993) and Cogley and Nason (1995). More recently, however, Baxter and King (1995) argue that HP filtering can produce reasonable approximations to an ideal business cycle filter. We therefore believe that the results reported in the next section are reasonably robust across business cycle filters.

12.3 Data and Results

We study monthly data (from Tick Data) on spot-month futures prices for crude oil, heating oil, unleaded gasoline and natural gas — spot-month futures prices are used as a proxy for current cash prices. Since these commodities began trading at different times on the New York Mercantile Exchange (NYMEX), we have a different sample size for each of these commodities. In particular, crude oil began trading in March 1983, heating oil in March 1979, unleaded gasoline in December 1984 and natural gas in April 1990. To investigate the cyclical behavior of energy prices, we match them with the U.S. industrial production index, consumer price level and unemployment rate, using data on these variables up to April 1993. This match produces 122 monthly observations for crude oil, 157 for heating oil, 94 for unleaded gasoline and 37 for natural gas.

Table 12.1 reports the contemporaneous and the cross correlations (at lags and leads of 1-6 months) between the cyclical components of energy prices and the cyclical component of U.S. industrial production (in panel A), the unemployment rate (in panel B) and consumer prices (in panel C). A number near 1 in the x_t column of panel A indicates strong procyclical movements and a number near -1 indicates strong countercyclical movements. The numbers in the remaining columns indicate the phase shift

TABLE 12.1

HP Cyclical Correlations of Spot-Month Energy Futures Prices With U.S. Output, Prices and the Unemployment Rate

Commodity	Correlation coefficients												
	x_{t-6}	x_{t-5}	x_{t-4}	x_{t-3}	x_{t-2}	x_{t-1}	x_t	x_{t+1}	x_{t+2}	x_{t+3}	x_{t+4}	x_{t+5}	x_{t+6}
A. Cross Correlations with U.S. Industrial Production													
Crude oil	-0.02	0.01	0.07	0.16	0.23	0.29	0.31	0.28	0.24	0.22	0.19	0.14	0.10
Heating oil	-0.05	0.01	0.06	0.11	0.17	0.24	0.29	0.29	0.29	0.27	0.20	0.15	0.11
Unleaded gasoline	-0.19	-0.13	-0.07	0.04	0.15	0.24	0.30	0.32	0.32	0.35	0.27	0.17	0.12
Natural gas	0.04	-0.02	-0.06	-0.02	0.06	0.19	0.34	0.46	0.53	0.51	0.43	0.25	0.01
B. Cross Correlations with the U.S. Unemployment Rate													
Crude oil	0.00	-0.04	-0.10	-0.16	-0.20	-0.23	-0.26	-0.26	-0.24	-0.26	-0.25	-0.22	-0.21
Heating oil	0.01	-0.05	-0.11	-0.16	-0.21	-0.25	-0.28	-0.28	-0.25	-0.25	-0.21	-0.17	-0.14
Unleaded gasoline	0.15	0.08	0.02	-0.04	-0.10	-0.17	-0.26	-0.29	-0.30	-0.35	-0.34	-0.30	-0.26
Natural gas	-0.22	-0.25	-0.19	-0.12	-0.17	-0.21	-0.29	-0.28	-0.28	-0.31	-0.37	-0.29	-0.11
C. Cross Correlations with U.S. Consumer Prices													
Crude oil	0.33	0.44	0.54	0.62	0.68	0.66	0.51	0.32	0.17	0.06	-0.02	-0.05	-0.05
Heating oil	0.17	0.22	0.30	0.36	0.39	0.37	0.28	0.18	0.10	0.08	0.04	0.02	0.04
Unleaded gasoline	0.29	0.39	0.53	0.64	0.72	0.69	0.55	0.40	0.34	0.26	0.15	0.05	-0.02
Natural gas	0.23	0.24	0.24	0.22	0.18	0.10	0.04	-0.04	-0.17	-0.36	-0.51	-0.58	-0.48

Note: Results are reported using monthly data for the following sample periods: crude oil, 1983:3-1993:4; heating oil, 1979:3-1993:4; unleaded gasoline, 1979:12-1993:4; and natural gas, 1990:4-1993:4.

TABLE 12.2

HP CYCLICAL CORRELATIONS OF SPOT-MONTH HEATING OIL, UNLEADED GASOLINE AND NATURAL GAS FUTURES PRICES WITH SPOT-MONTH CRUDE OIL FUTURES PRICES

Commodity	Correlation coefficients												
	x_{t-6}	x_{t-5}	x_{t-4}	x_{t-3}	x_{t-2}	x_{t-1}	x_t	x_{t+1}	x_{t+2}	x_{t+3}	x_{t+4}	x_{t+5}	x_{t+6}
Heating oil	-0.14	-0.14	-0.06	0.09	0.37	0.66	0.88	0.76	0.60	0.42	0.27	0.09	0.01
Unleaded gasoline	-0.10	0.00	0.13	0.23	0.47	0.72	0.86	0.66	0.42	0.29	0.25	0.16	0.09
Natural gas	-0.18	-0.28	-0.38	-0.35	-0.18	0.09	0.33	0.38	0.27	0.03	-0.22	-0.33	-0.37

Note: Results are reported using monthly data for the following sample periods: heating oil, 1979:3-1993:4; unleaded gasoline, 1979:3-1993:4; and natural gas, 1990:4-1993:4.

relative to industrial production. For example, a series that leads (lags) the cycle by 3 months will have its maximum value in the x_{t-3} (x_{t+3}) column.

As panel A of Table 12.1 shows, energy prices are weakly procyclical, with natural gas prices being more so. Moreover, the cycles of crude oil and heating oil prices coincide with the industrial production cycle, while those of unleaded gasoline and natural gas lag the cycle of industrial production. This has important implications for hedgers and speculators. If speculators, for example, expect an increase in real output, they may wish to buy futures since the price of energy commodities is likely to rise.

To investigate the robustness of these results to changes in the cyclical indicator, we report in panel B of Table 12.1 correlations (in the same fashion as in panel A) using the unemployment rate as the cyclical indicator. Of course, since the cyclical component of industrial production and the unemployment rate are negatively correlated, a negative correlation in panel B indicates procyclical variation and a positive correlation indicates countercyclical variation. Clearly, the results of panel B in general confirm those in panel A. Hence, we conclude that irrespective of the cyclical indicator, energy prices are procyclical.

Panel C of Table 12.1 shows cyclical energy prices-U.S. consumer prices correlations. Clearly, crude oil and unleaded gasoline prices are strongly contemporaneously correlated with U.S. consumer prices, while heating oil prices are weakly correlated and natural gas prices are independent. Moreover, the cycles of crude oil, heating oil and unleaded gas prices lead the cycle of U.S. consumer prices, suggesting that changes in energy prices might be good predictors of future aggregate price changes. This also raises the possibility that energy prices might be a useful guide for monetary policy, possibly serving as an important indicator variable.

Finally, in Table 12.2 we show HP cyclical correlations of heating oil, unleaded gasoline and natural gas prices with crude oil prices. The results indicate that the contemporaneous correlations are strikingly strong in the case of heating oil and unleaded gasoline but not as strong in the case of natural gas. This is consistent with the conclusion reached by Serletis (1994) that crude oil, heating oil and unleaded gasoline prices are driven by one common trend, suggesting that it is appropriate to model these prices as a cointegrated system. Natural gas prices, however, seem to react to a separate set of fundamentals.

12.4   Conclusion

In this chapter we investigated the cyclical behavior of energy prices using monthly data and the methodology suggested by Kydland and Prescott

(1990). Based on stationary HP cyclical deviations, our results are robust to alternative measures of the cycle and indicate that crude oil and heating oil prices are synchronous and procyclical whereas unleaded gasoline and natural gas prices are lagging procyclically. Moreover, energy prices are positively contemporaneously correlated with consumer prices and their cycles lead the cycle of consumer prices, suggesting a possible role for energy prices in the conduct of monetary policy.

However, the apparent phase-shift between energy prices and consumer prices should not be interpreted as supporting an effect from energy prices to consumer prices since using lead-lag relationships to justify causality is tenuous. Clearly, the investigation of the empirical relationship between energy prices and consumer prices, by looking at the performance of energy prices as indicators of inflation, is an area for potentially productive future research. Such an examination could utilize current state-of-the-art econometric methodology, such as, for example, integration and cointegration theory as well as error-correction modeling (if applicable), using either the single-equation approach of a multi-equation (VAR) framework.

We also presented evidence regarding cyclical correlations of heating oil, unleaded gasoline and natural gas prices with crude oil prices. We show that the contemporaneous crude oil-heating oil and crude oil-unleaded gasoline correlations are very strong, providing future support to the conclusion of Serletis (1994) that these prices are driven by only one common trend which means, according to the interpretation of Stock and Watson (1988), that the same underlying stochastic components presumably affect the crude oil, heating oil and unleaded gasoline markets. The natural gas market, however, doesn't seem to be linked to the crude oil market.

The results presented in this chapter pertain to the United States. Of course, the cyclical behavior of energy prices in countries with different industrial structures and/or levels of oil dependency would be expected to be different. Therefore the international generalizability of this work is also an area for future research.

Chapter 13

The Message in North American Energy Prices

*Apostolos Serletis and John Herbert**

13.1 Introduction

In the last decade, the natural gas industry has seen a dramatic transformation from a highly regulated industry to one which is more market-oriented. The transition to a less regulated, more market-driven environment has significantly affected business operations. In particular, production sites, pipelines, and transmission and storage services are more accessible today, thereby ensuring that changes in market demand and supply are reflected in prices on spot, futures, and swaps markets. There is also a dynamic power industry in North America, the dynamics of which cannot be captured by any given relationship to crude oil or natural gas. They seem to be driven by the variety and seasonality of applications.

In this chapter, we investigate the dynamics of natural gas, fuel oil, and power prices in the mid-Atlantic area of eastern Pennsylvania, New Jersey, Maryland, and Delaware (an area in which as much oil, natural gas, and power is used as in all of Britain). These prices are expected to be related for several reasons. Fuel oil and natural gas, for example, are used as substitutes in industrial boiler, and oil and natural gas are used as peaking sources of supply for power generation for cooling loads in the summer and for heating loads in the winter. Moreover, all these types of energy directly

---

*Originally published in *Energy Economics* 21 (1999), 471-483. Reprinted with permission.

serve space heating demands during the winter. Thus, wholesale prices for these sources of energy are expected to respond similarly to different types of shocks.

In investigating whether such key North American natural gas, power, and fuel oil markets are linked together, we test for shared price trends. In doing so, following King and Cuc (1996) and Serletis (1997), we use current state-of-the-art econometric methodology. In particular, we pay explicit attention to whether or not the variables are stationary. It is an empirical fact that many important macroeconomic and financial variables appear to be integrated. If the series are integrated, but not cointegrated, ordinary least squares yields misleading results. Under these circumstances it becomes important to evaluate empirically the time series properties of the variables and to test for cointegration.

The chapter is organized along the following lines. Section 13.2 discusses the data and provides some graphical representations. Sections 13.3 and 13.4 investigate the integration and cointegration properties of the price series and interpret the results in terms of convergence and the existence of unexploited profit opportunities. Section 13.5 tests for Granger causality, explicitly taking into account the univariate and bivariate time series properties of the variables. The last section concludes with some suggestions for potentially productive future empirical research.

13.2 Some Basic Facts

We use daily data from 25/10/96 to 21/11/97 on the Henry Hub and Transco-Zone 6 natural gas prices - the Henry Hub natural gas price is strongly correlated with the New York Mercantile Exchange (NYMEX) Henry Hub spot month futures price, while Transco Zone 6 is an important segment of the Transco pipeline extending from Northern Virginia to New York City, serving the eastern seaboard. We also use the Pennsylvania, New Jersey, Maryland (PJM) power market for electricity prices, over the same time period and frequency. This power market serves the same general area as Transco-Zone 6 and has regularly been considered as a delivery point for a power futures contract. Finally, we use fuel oil prices for New York Harbor which is the delivery point for the NYMEX heating oil contract - it is also a standard reference price for oil in the Northeast.

One interesting feature of the data is the contemporaneous correlation between the different price series. These correlations are reported in Table 13.1 for log levels (in panel A) and for first differences of log levels (in panel B). To determine whether these correlations are statistically significant, Pindyck and Rotemberg (1990) is followed and a likelihood ratio test of the

TABLE 13.1
CONTEMPORANEOUS CORRELATIONS BETWEEN PRICES

	A. Log Levels				B. First differences of log levels			
	Henry hub	Transco zone 6	Power	Fuel oil	Henry hub	Transco zone 6	Power	Fuel oil
Henry hub	1				1			
Transco zone 6	0.962	1			0.528	1		
PJM power	0.196	0.207	1		0.081	0.173	1	
Fuel oil	0.611	0.716	0.013	1	0.068	0.038	0.060	1
	$\chi^2(6) = 984.21$				$\chi^2(6) = 99.11$			

Note: Daily data, 25 October 1996 to 21 November 1997.

hypotheses that the correlation matrices are equal to the identity matrix is performed. The test statistic is

$$-2\ln(|R|N/2)$$

where $|R|$ is the determinant of the correlation matrix and N is the number of observations. This test statistic is distributed as χ^2 with $0.5q(q-1)$ degrees of freedom, where q is the number of series.

The test statistic is 984.21 with a p-value of 0.000 for the logged prices and 99.11 with a p-value of 0.000 for the first-differenced logged prices. Clearly, the hypothesis that these price are uncorrelated is rejected. Notice, however, that the correlations indicate a lack of a relationship between power and the other series. The correlation patterns documented in Table 13.1 manifest in the graphical representation of the series in Figure 13.1, for logged levels.

13.3 The Integration Properties of the Variables

The first step in examining trends in a set of variables is to test for the presence of a stochastic trend (a unit root) in the autoregressive representation of each individual series. Nelson and Plosser (1982) argue that most macroeconomic and financial time series have a unit root (a stochastic trend), and describe this property as one of being 'difference stationary' (DS) so that the first difference of a time series is stationary. An alternative 'trend stationary' model (TS) has been found to be less appropriate.

In what follows we test the null hypothesis of a stochastic trend against the trend-stationary alternative by estimating by ordinary least-squares (OLS) the following augmented Dickey-Fuller (ADF) type regression [see Dickey and Fuller (1981)]

$$\Delta z_t = a_0 + a_2 t + \gamma z_{t-1} + \sum_{j=1}^{k} b_j \Delta z_{t-j} + \varepsilon_t \qquad (13.1)$$

where Δ is the difference operator such that $\Delta z_t = z_t - z_{t-1}$. The k extra regressors in (13.1) are added to eliminate possible nuisance parameter dependencies in the limit distributions of the test statistics caused by temporal dependencies in the disturbances. The optimal lag length (that is, k) is taken to be the one selected by the Akaike information criterion (AIC) plus 2 — see Pantula et al. (1994) for details regarding the advantages of this rule for choosing the number of augmenting lags in equation (13.1).

Table 13.2 presents the results. The first column of Table 13.2 gives the optimal value of k in equation (13.1), based on the AIC plus 2 rule, for each price series. This identifies k to be 4 for the Henry Hub natural gas price series, 2 for the Transco Zone 6 natural gas price series and the fuel oil price series, and 3 for the power price series.

The t-statistics for the null hypothesis $\gamma = 0$ in equation (13.1) are given under τ_τ in Table 13.2. Under the null hypothesis that $\gamma = 0$, the appropriate critical value of τ_τ at the 5% level (with 200 observations) is -3.45 — see Fuller (1976, Table 8.5.2). Hence, the null hypothesis of a unit root is rejected only in the case of the power price. For this series, we conclude at this stage that it does not contain a unit root [or in the terminology of Engle and Granger (1987) that it is $I(0)$].

For the remaining series, for which the null hypothesis of a unit root has not been rejected, there is a question concerning the test's power in the presence of the deterministic part of the regression (i.e., $a_0 + a_2 t$). In particular, one problem is that the presence of the additional estimated parameters reduces degrees of freedom and the power of the test — reduced power means that we will conclude that the process contains a unit root when, in fact, none is present. Another problem is that the appropriate statistic for testing $\gamma = 0$ depends on which regressors are included in the model.

Although we can never be sure of the actual data-generating process, here we follow the procedure suggested by Dolado *et al.* (1990) for testing for a unit root when the form of the data-generating process is unknown. In particular, since the null hypothesis of a unit root is not rejected, it is necessary to determine whether too many deterministic regressors are included in equation (13.1).We therefore test for the significance of the trend term in equation (13.1) under the null of a unit root, using the $t(a_2)$ statistic in Table 13.2. Under the null that $a_2 = 0$ given the presence of a unit root, the appropriate critical value of $t(a_2)$ at the 5% significance level is 2.79 — see Dickey and Fuller (1981). Clearly, the null cannot be rejected, suggesting that the trend is not significant. The ϕ_3 statistic which tests the joint null hypothesis $a_2 = \gamma = 0$ reconfirms this result.

This means that we should estimate the model without the trend, i.e., in the following form

$$\Delta z_t = a_0 + \gamma z_{t-1} + \sum_{j=1}^{k} b_j \Delta z_{t-j} + \varepsilon_t \qquad (13.2)$$

and test for the presence of a unit root using the τ_μ statistic. The results, reported in Table 13.2, indicate that the null hypothesis of a unit root

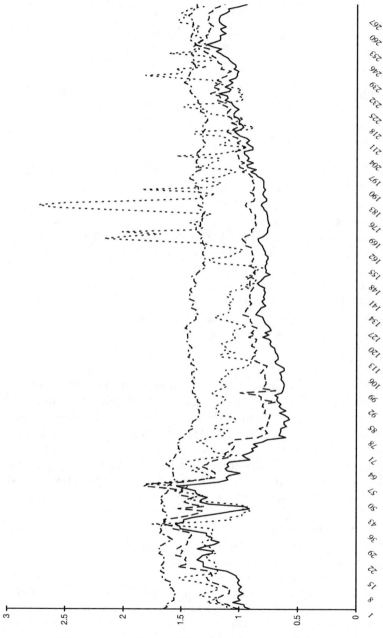

Figure 13.1: Logged North American energy prices

TABLE 13.2

UNIT ROOT TEST RESULTS

Series	k	τ_τ	$t(a_2)$	ϕ_3	τ_μ	Decision
			Test statistics			
Henry hub	4	-2.03	-0.14	2.12	-2.03	I(1)
Transco zone 6	2	-1.45	0.04	1.17	-1.51	I(1)
PJM power	3	4.61*	0.73	10.94*	-4.56*	I(0)
Fuel oil	2	-1.81	-1.15	1.73	-1.43	I(1)

Notes: Daily data, 25 October 1996 - 21 November 1997. All the series are in logs. An asterisk indicates rejection of the null hypothesis at the 5% significance level. τ_τ is the t-statistic for the null hypothesis $\gamma = 0$ in Eq. (13.1). Under the null hypothesis, the appropriate critical value of τ_τ at the 5% significance level (with 100 observations) is -3.45 — see Fuller (1976, Table 8.5.2). $t(a_2)$ is the t-statistic for the presence of the time trend (i.e. the null hypothesis $a_2 = 0$) in Eq. (13.1), given the presence of a unit root. The appropriate 95% critical value for $t(a_2)$, given by Dickey and Fuller (1981), is 2.79. The ϕ_3 statistic tests the joint null $a_2 = \gamma = 0$ in Eq. (13.1). The 95% critical value, given by Dickey and Fuller (1981) is 6.49. Finally, τ_μ is the t-statistic for the null $\gamma = 0$ in Eq. (13.2). The appropriate 95% critical value of τ_μ is -2.89 — see Fuller (1976, Table 8.5.2).

cannot be rejected for the Henry Hub and Transco Zone 6 natural gas price series as well as for the fuel oil price series. Our decision regarding the univariate time series properties of these series is summarized in the last column of Table 13.2. Intuitively, fluctuations in a stationary series are viewed as temporary deviations from its underlying trend and are expected to return to its (more or less constant) trend growth rate in the long run. In the case, however, of integrated series, such deviations should be treated as permanent — that is, there is no tendency for the series to revert to its deterministic path.

Our results regarding the univariate time series properties of the variables are also useful in regard to the decision of whether to specify univariate models [such as, for example, moving-average (MA) models, autoregressive (AR) models, and autoregressive moving-average (ARMA) models] in levels or first differences. If the series are stationary (i.e., there is no unit root), then it is desirable to work in levels, and if the series are integrated (i.e., there is a unit root), then differencing is appropriate.

In a regression analysis context, however, the appropriate way to treat integrated variables is not so straightforward. It is possible, for example, that the integrated variables 'cointegrate' — in the sense that a linear relationship among the variables is stationary. Differencing such an already stationary relationship entails a misspecification error, which we should avoid. It is to this issue that the next section is devoted.

13.4 Shared Price Trends

Since a stochastic trend has been confirmed for the natural gas and fuel price series, we now explore for shared stochastic trends among these series using methods recommended by Engle and Granger (1987). In doing so, we test for cointegration (i.e., long-run equilibrium relationships). Cointegration is a relatively new statistical concept designed to deal explicitly with the analysis of the relationship between nonstationary time series. In particular, it allows individual time series to be nonstationary, but requires a linear combination of the series to be stationary. Therefore, the basic idea behind cointegration is to search for a linear combination of individually nonstationary time series that is itself stationary. Evidence to the contrary provides strong empirical support for the hypothesis that the integrated variables have no inherent tendency to move together over time.

Consider, for example, the null hypothesis that there is no cointegration between two price series y_t and x_t [or equivalently, there are no shared stochastic trends (i.e., there are two distinct trends) between these series, in the terminology of Stock and Watson (1988)]. The alternative hypothesis is that there is cointegration (or equivalently, they share a stochastic trend). Following Engle and Granger (1987), we estimate the so-called cointegrating regression (selecting arbitrarily a normalization)

$$y_t = \alpha + \beta x_t + \varepsilon_t \tag{13.3}$$

where ε_t denotes the OLS regression residuals. A test of the null hypothesis of no cointegration (against the alternative of cointegration) is based on testing for a unit root in the regression residuals ε_t using the ADF test and simulated critical values which correctly take into account the number of variables in the cointegrating regression.

Table 13.3 shows marginal significance levels for Engle-Granger cointegration tests between the integrated price series. Clearly, the null hypothesis of no cointegration (i.e., absence of shared stochastic trends) is rejected (at the 1% significance level). It is to be noted that these results are robust to the selected normalization. Under the common trends interpretation [see, for example, Stock and Watson (1988)] these results are not too surprising. The same underlying stochastic growth components presumably affect all three markets, implying that these three markets are linked together, with the power market, of course, being segmented. Notice that since the power price series is $I(0)$ and each of the other price series is $I(1)$, inferences regarding the strength of the relationship between the power price series and each of the other price series will be spurious.

One way to interpret these results is in terms of the absence or presence of unexploited profit opportunities. In the case, for example, of integrated price series that do not cointegrate, the difference between the respective prices fluctuates stochastically, in excess of transmission and transaction costs, indicating the failure of potential arbitrage to discipline prices. In this case, the marginal value of the commodity across locations would differ by more than transmission and transaction costs suggesting unexploited profit opportunities. In the case, however, of integrated price series that cointegrate, the price differential is stationary, implying price convergence, a high degree of price competition, and the absence of unexploited profit opportunities. In this case, every permanent shock in the trend of one series is ultimately transmitted to the trend of the other series.

TABLE 13.3

MARGINAL SIGNIFICANCE LEVELS OF
ENGLE AND GRANGER (1987) COINTEGRATION TESTS
FOR THOSE PRICE SERIES THAT ARE INTEGRATED

	Transco zone 6		*Fuel oil*	
	k	p-value	k	p-value
Henry hub	8	0.001	3	0.006
Transco zone 6			2	0.001

Notes: Daily data, 25 October 1996-21 November 1997. The null hypothesis is the absence of cointegration. Low p-values imply strong evidence against the null. The dependent variable in the cointegrating regression is the one indicated in the row heading — the results are robust to this normalization.

In fact, for the price pairs that we have established that they share a stochastic trend, the factors of proportionality for shared stochastic trends [the β's in equation (13.3)] can be consistently estimated using ordinary least squares [see, for example, Stock (1987)]. These are reported in Table 13.4. Let us consider the relationship between the Henry Hub-Transco Zone 6 natural gas price pair, reported in Table 13.4. Clearly, this is a statistically significant relationship and, in particular, a 1% increase in the Transco Zone 6 natural gas price is associated with a 0.915 percentage point

increase in the Henry Hub natural gas price. The remaining numbers in
Table 13.4 should be interpreted along these lines.

TABLE 13.4

ESTIMATES OF FACTORS OF PROPORTIONATLITY
FOR SHARED STOCHASTIC TRENDS FOR THOSE
PRICE PAIRS THAT COINTEGRATE

	Transco zone 6			Fuel oil		
	β	$t(\beta)$	R^2	β	$t(\beta)$	R^2
Henry hub	0.915	58.5	09.26	1.179	12.7	0.374
Transco zone 6				1.452	16.8	0.513

Notes: Daily data, 25 October 1996-21 November 1997. Estimates of factors
of proportionatliy for shared stochastic trends (based on logarithmically transformed
price series) are reported for only those price pairs that cointegrate.

What is key for any firm is whether differences between prices at different locations, such as Hennry Hub and Transco-Zone 6, exceed the cost of
making trading arrangements between these locations which may or may
not involve the actual movement of gas between these locations. Such arrangements may, for example, involve the movement of gas out of storage
in the Northeast and later replacement of this gas with gas from Louisiana.
Thus, methods of trading gas 'between locations' vary greatly between companies because of differences in asset and contract mix. Moreover, superior
knowledge of trading conditions at a large number of locations will allow
firms with operationally flexible assets to exploit differences between prices
at different locations.

Of course, the combination of inflexible transportation contracts, regulation, and poor information on available transportation and storage capacity preclude many firms from exploiting profit opportunities, whenever
they arise. However, unregulated firms with superior knowledge of capacity availability and flexible, active contracting programs regularly exploit
such opportunities. A still significant number of regulated firms in an increasingly deregulated industry allows the less regulated firms to exploit
opportunities created by the relatively inflexible business and operating
practices of the regulated part.

13.5   Error Correction Estimates and Causality Tests

If two series cointegrate, there is a long-run relationship between them. Moreover, according to the Granger representation theorem, the short-run dynamics can be described by the error correction model (ECM). In an error correction model, the short-term dynamics of the variables in the system are influenced by the deviation from long-run equilibrium. In other words, if the system is to return to the long-run equilibrium, the movements of at least some of the variables must be influenced by the magnitude of the deviation from the long-run relationship. If, for example, the gap between two cointegrating natural gas price series, y_t and x_t, is large relative to the long-run relationship, the gap must ultimately close by adjustments in y_t, x_t, or both.

If the y_t and x_t series are cointegrated, the residual ε_t in equation (13.3) estimates the deviation from long-run equilibrium in period $t - 1$, and can be used to estimate the error-correction model, which Engle and Granger (1987) argue will have the following form

$$\Delta y_t = \alpha_1 + \alpha_y \hat{\varepsilon}_{t-1} + \sum_{j=1}^{r} \alpha_{11}(j)\Delta y_{t-j} + \sum_{j=1}^{s} \alpha_{12}(j)\Delta x_{t-j} + \varepsilon_{yt} \quad (13.4)$$

$$\Delta x_t = \alpha_2 + \alpha_x \hat{\varepsilon}_{t-1} + \sum_{j=1}^{r} \alpha_{21}(j)\Delta y_{t-j} + \sum_{j=1}^{s} \alpha_{22}(j)\Delta x_{t-j} + \varepsilon_{xt} \quad (13.5)$$

This is a bivariate vector autoregression (VAR) in first differences, augmented by the error-correction term, ε_t. The error-correction model clearly shows how y_t and x_t change in response to stochastic shocks (represented by ε_{yt} and ε_{xt}) and to the previous period's deviation from long-run equilibrium (represented by ε_{t-1}).

If, for example, ε_{t-1} is positive (so that $y_{t-1} - \alpha - \beta x_{t-1} > 0$), x_t would rise and y_t would fall until long-run equilibrium is attained, when $y_t = \alpha + \beta x_t$. Notice that α_y and α_x can be interpreted as speed of adjustment parameters. For example, the larger is α_y, the greater the response of y_t to the previous period's deviation from long-run equilibrium. On the other hand, very small values of α_y imply that y_t is unresponsive to last period's equilibrium error. In fact, for Δy_t to be unaffected by x_t, y_t and all the $\alpha_{12}(j)$ coefficients in (13.4) must be equal to zero. This is the empirical definition of Granger causality in cointegrated systems. In other words, the absence of Granger causality for cointegrated variables requires the additional condition that the speed of adjustment coefficient be equal to zero.

Thus, one could determine the causal relationship between y_t and x_t by first fitting equation (13.4) by ordinary least squares and obtaining the unrestricted sum of squared residuals, SSR_u. Then by running another regression equation under the null hypothesis that α_y and all the coefficients of the lagged values of Δx_t are zero, the restricted sum of squared residuals, SSR_r, is obtained. The statistic

$$\frac{(SSR_r - SSR_u)/(s+1)}{SSR_u/(T-r-s-2)}$$

has an asymptotic F-distribution with numerator degrees of freedom $(s+1)$ and denominator degrees of freedom $(T-r-s-2)$, where T is the number of observations, s represents the number of lags for Δx_t in equation (13.4), and 2 is subtracted out to account for the constant term and the error correction term in equation (13.4). If the null hypothesis cannot be rejected, than the conclusion is that the data do not show causality. If the null hypothesis is rejected, then the conclusion is that the data do show causality. The roles of y_t and x_t are reversed in another F test [as in equation (13.5)] to see whether there is a feedback relationship among these series.

One preliminary matter also had to be dealt with before we could proceed to estimate the error-correction model and perform Granger-causality tests. It concerns the lengths of lags r and s in equations (13.4) and (13.5). In the literature r and s are frequently chosen to have the same value, and lag lengths of 4, 6, or 8 are used most often. Such arbitrary lag specifications can produce misleading results, however, because they may imply misspecification of the order of the autoregressive process. For example, if either r or s (or both) is too large, the estimates will be unbiased but inefficient. If either r or s (or both) is too small, the estimates will be biased but have a smaller variance.

Here, we used the data to determine the 'optimum' lag structure. In particular, the optimal r and s in each of equations (13.4) and (13.5) was determined using Akaike's information criterion (AIC). The AIC was calculated as

$$AIC(r,s) = \log\left(\frac{SSR}{T}\right) + 2\left(\frac{r+s+1}{T}\right)$$

where T is the number of observations and SSR is the sum of squared residuals. Note that the AIC balances the degrees of freedom used (as implied by the second term in the expression) and the fit of the equation (as implied by SSR).

We used the AIC with a maximum value of 12 for each of r and s in equations (13.4) and (13.5) and by running 144 regressions for each bivariate relationship we chose the one that produced the smallest value for the AIC. Based on these optimal specifications, in Tables 13.5 and 13.6 we present estimates of the speed of adjustment parameters (along with t-ratios) as well as p-values for Granger causality F-tests (for those price series that cointegrate). The signs of the speed of adjustment coefficients are in accord with convergence toward the long-run equilibrium — that is, the absolute values of the speed of adjustment coefficients are not too large. The numbers apply to an error-correction model in which the left-hand-side variable is the one indicated in the row heading.

Consider the Henry Hub - Transco Zone 6 natural gas price relationship. With the Henry Hub natural gas price as the dependent variable in equation (13.4), the estimated speed of adjustment coefficient (α_y) is 0.026 with a t-ratio of 0.35, indicating that it is not significant. When, however, the Transco Zone 6 price is used as the dependent variable in equation (13.5), the results in Table 13.6 indicate that the estimated speed of adjustment coefficient α_x is -0.161 and significant (the t-ratio is -2.36). This means that the Transco Zone 6 natural gas price tends to decrease significantly in response to a positive discrepancy between the Henry Hub price and the Transco Zone 6 price in the previous period. Clearly, this is useful information for a trading company regarding the design of a successful trading strategy.

Finally, the p-value of 0.001 in Table 13.5 indicates that the null hypothesis that $\alpha_y = \alpha_{12}(1) = \alpha_{12}(11) = 0$ in equation (13.4) is rejected, implying that Transco Zone 6 natural gas prices do Granger cause Henry Hub prices. Also, the p-value of .001 in Table 13.6 indicates that the null $\alpha_x = \alpha_{22}(1) = \cdots = \alpha_{22}(12) = 0$ in equation (13.5) is rejected, implying that Henry Hub prices Granger cause Transco Zone 6 prices. In other words, knowledge of past Henry Hub prices improves the prediction of future Transco Zone 6 prices beyond predictions that are based on past Transco Zone 6 prices alone. It should be noted that Granger causality refers merely to predictability and has no implications for the strength of conclusions which refer to underlying structural factors.

13.6   Conclusion

The chapter tested for unit roots in the univariate time-series representation of daily Henry Hub and Transco Zone 6 natural gas prices, as well as of power and fuel prices. Based on augmented Dickey-Fuller (ADF) unit root testing procedures, the results show that the random-walk hypothesis cannot be rejected for the natural gas and fuel oil prices. The power price series, however, appears to be stationary. The implications of these findings

TABLE 13.5

ESTIMATED SPEED OF ADJUSTMENT PARAMETERS AND
MARGINAL SIGNIFICANCE LEVELS FOR GRANGER CAUSALITY TESTS
FOR THOSE PRICE PAIRS THAT COINTEGRATE

$$\Delta y_t = \alpha_1 + \alpha_y \hat{\varepsilon}_{t-1} + \sum_{j=1}^{r} \alpha_{11}(j) \Delta y_{t-j} + \sum_{j=1}^{s} \alpha_{12}(j) \, \Delta x_{t-j} + \hat{\varepsilon}_{yt}$$

	Transco zone 6				Fuel oil		
	(r, s)	α_y	$t(\alpha_y)$	p-value	(r, s)	α_y	$t(\alpha_y)$	p-value
Henry hub	(12,11)	0.026	0.35	0.001	(1,1)	-0.042	-2.50	0.223
Transco zone 6					(1,3)	-0.047	-2.28	0.097

Notes: Daily data, 25 October 1996-21 November 1997. The dependent variable is the one indicated in the row heading. Numbers in parenthesis indicate the optimal (in the minimum AIC sense) lag specification. p-values less than 0.05 reject the null hypothesis of no causality at the 0.05 level of significance.

TABLE 13.6
ESTIMATED SPEED OF ADJUSTMENT PARAMETERS AND MARGINAL SIGNIFICANCE LEVELS FOR REVERSE GRANGER CAUSALITY TESTS FOR THOSE PRICE PAIRS THAT COINTEGRATE

$$\Delta x_t = \alpha_2 + \alpha_x \hat{\varepsilon}_{t-1} + \sum_{j=1}^{r} \alpha_{21}(j)\Delta x_{t-j} + \sum_{j=1}^{s} \alpha_{22}(j)\Delta y_{t-j} + \hat{\varepsilon}_{xt}$$

	Transco zone 6				Fuel oil			
	(r,s)	α_x	$t(\alpha_x)$	p-value	(r,s)	α_x	$t(\alpha_x)$	p-value
Transco zone 6	(11,12)	-0.161	-2.36	0.001				
Fuel oil	(1,1)	-0.029	-2.21	0.331	(1,1)	-0.035	-2.36	0.294

Notes: Daily data, 25 October 1996-21 November 1997. The dependent variable is the one indicated in the row heading. Numbers within parenthesis indicate the optimal (in the minimum AIC sense) lag specification. p-values less than 0.05 reject the null hypothesis of no causality at the 0.05 level of significance.

regarding the long-run effect (or persistence) of a shock on the level of these series were also discussed. It was argued, for example, that shocks to an integrated series are permanent and to a stationary series temporary.

Moreover, the application of Engle and Granger (1987) cointegration methods to explore the degree of shared trends (for those series for which a stochastic trend has been confirmed), revealed that there are shared trends among the Henry Hub and Transco Zone 6 natural gas prices and the fuel oil price. This means that there are empirically effective arbitraging mechanisms for these prices across these markets. The estimation of error-correcting causality models for the integrated price series also revealed causality and a feedback relationship between any two price pairs.

We have used univariate, and bivariate models to draw valid inferences about the time series relations between energy prices. Alternative and perhaps more general and more robust specifications could be estimated. A particularly constructive approach would be based on the use of higher-dimensional VARs. Impulse response functions and variance decompositions are the hallmark of VAR analysis focusing on higher-order VARs is an area for potentially productive future research.

Chapter 14

Testing for Common Features in North American Energy Markets

*Apostolos Serletis and Ricardo Rangel-Ruiz**

14.1 Introduction

In recent years, the North American energy industry has undergone major structural changes that have significantly affected the environment in which producers, transmission companies, utilities and industrial customers operate and make decisions. For example, major policy changes are the U.S. Natural Gas Policy Act of 1978, Natural Gas Decontrol Act of 1989, and FERC Orders 486 and 636. In Canada, deregulation in the mid-1980s has also broken the explicit link between the delivered prices of natural gas and crude oil (that was in place prior to 1985), and has fundamentally changed the environment in which the Canadian oil and gas industry operates. Moreover, the Free Trade Agreement (FTA) signed in 1988 by the United States and Canada, and its successor, the North American Free Trade Agreement (NAFTA) signed in 1993 by the United States, Canada, and Mexico, have underpinned the process of deregulation and attempted to increase the efficiency of the North American energy industry.

The main objective of this chapter is to assess the strength of shared dy-

*Originally published in *Energy Economics* 26 (2004), 401-414. Reprinted with permission.

namics between North American energy markets in the period after deregulation. In doing so, we provide a first look at shared trends and shared cycles between the West Texas Intermediate (WTI) crude oil and Henry Hub natural gas markets, drawing on recent developments on cointegration theory. We are interested in whether the link between these two markets weakened in the deregulated period, as competition and market forces played a greater role in determining prices. Moreover, we explore the inter-connectedness of North American energy markets by investigating the strength of shared features between the U.S. Henry Hub and AECO Alberta natural gas prices. We are interested in whether Canadian export prices to the United States are simply linear transformations of the U.S. Henry Hub price.

Shared stochastic trends between different energy markets have been investigated in a number of recent studies — see, example, Serletis (1994), Serletis and Herbert (1999), and Plourde and Watkins (2000). These studies, however, typically require the researcher to take a stance on a common order of integration for the individual price series. As a result, most of the literature ignores a recent important contribution to this topic by Ng and Perron (1997) who show that we should be wary of estimation and inference in nearly unbalanced nearly cointegrated systems. In this chapter we use the recent Pesaran *et al.* (2001) bounds testing approach to the investigation of long run relationships. This is a particularly relevant methodology as it does not require that we take a stand on the time series properties of the data. Therefore we are able to test for the existence of a long-run relationship without having to assume that the series are integrated of order zero [or I(0) in the terminology of Engle and Granger (1987)] or I(1).

Our principal concern, however, is with the dynamics of North American natural gas and crude oil markets. The distinctive feature of our contribution is that we test for shared cycles (and when appropriate for codependent cycles) using the recently developed testing procedures by Engle and Kozicki (1993) and Vahid and Engle (1993). Our main objective is to determine the strength of the dynamic relationship between natural gas and crude oil markets, judged according to whether they respond in a similar manner to cycle generating innovations. The Engle and Kozicki (1993) and Vahid and Engle (1993) approach provides a stronger and more informative test of cyclical comovements than the previously used [by Serletis and Kemp (1998)] Hodrick-Prescott (HP) contemporaneous and cross-correlation analysis.

The chapter is organized along the following lines. Section 14.2 reviews some basic theoretical results and relates them to the sharing of trends and cycles. Section 14.3 discusses the data and tests for common trends, cycles, and (where appropriate) codependent cycles in U.S. natural gas and crude oil markets. Section 14.4 investigates the inter-connectedness of

North American energy markets, and the last section briefly summarizes and concludes.

14.2 Common Trends and Common Cycles

Consider two variables y_t and x_t for which there may be possible long-run and/or short-run relationships. Following Stock and Watson (1988), we can decompose each variable into a trend, cyclical, and stationary (but not necessarily white-noise) irregular component as follows

$$y_t = \tau_{yt} + c_{yt} + \epsilon_{yt} \tag{14.1}$$

$$x_t = \tau_{xt} + c_{xt} + \epsilon_{xt}, \tag{14.2}$$

where τ_{jt} is the trend component of variable j at time t, c_{jt} is the cyclical component, and ϵ_{jt} is the noise (or irregular) component. In what follows we highlight some important differences between the traditional analysis of comovement and the more recent common cycles analysis.

14.2.1 Common Trends

If the individual series have a stochastic trend, we can explore for shared stochastic trends between the series. In particular, if the stochastic trend of x_t is shared with the y_t series (i.e., τ_{xt} is linearly related to τ_{yt}), then we have the following structure

$$y_t = \tau_{yt} + c_{yt} + \epsilon_{yt} \tag{14.3}$$

$$x_t = \alpha\tau_{yt} + c_{xt} + \epsilon_{xt} \tag{14.4}$$

where α is the factor of proportionality between the two trends. In this case there is a unique coefficient λ, such that the following linear combination of y_t and x_t

$$z_t = y_t - \lambda x_t$$

is a stationary series — see Engle and Granger (1987). In fact, if there is a shared stochastic trend, the linear combination z_t can be written as

$$z_t = \tau_{yt} + c_{yt} + \epsilon_{yt} - \lambda\left(\alpha\tau_{yt} + c_{xt} + \epsilon_{xt}\right)$$

$$= \tau_{yt} - \lambda\alpha\tau_{yt} + c_{yt} - \lambda c_{xt} + \epsilon_{yt} - \lambda\epsilon_{xt},$$

which for $\lambda = 1/\alpha$ reduces to

$$z_t = c_{yt} - \lambda c_{xt} + \epsilon_{yt} - \lambda \epsilon_{xt}.$$

Of course, λ may not be known a priori. Stock (1987) shows that λ can be consistently estimated using Ordinary Least Squares (OLS) in the following regression

$$y_t = \lambda x_t + z_t$$

The test for a common stochastic trend is therefore a cointegration test. That is, we test whether there is a cointegrating vector $[1, \lambda]$ such that z_t is stationary — see Engle and Granger (1987) for more details.

14.2.2 Common Cycles

Regarding common cycles, the approach adopted in the business cycle literature is a modern counterpart of the methods developed by Burns and Mitchell (1946). It involves the measurement of the degree of comovement between two series by the magnitude of the correlation coefficient, $\rho(j)$, $j \in \{0, \pm 1, \pm 2, \ldots\}$, between (stationary) cyclical deviations from trends. In particular, the contemporaneous correlation coefficient — $\rho(0)$ — gives information on the degree of contemporaneous comovement whereas the cross-correlation coefficient — $\rho(j)$ $j \in \{\pm 1, \pm 2, \ldots\}$ — gives information on the phase shift of one series relative to another — see Kydland and Prescott (1990) for details regarding the methodology and Serletis and Kemp (1998) for an application to energy markets.

An alternative more informative test for common cycles has recently been suggested by Engle and Kozicki (1993) and Vahid and Engle (1993) and is based on an extension of the common trends (cointegration) analysis in a stationary setting. They show that the presence of a cyclical component in the first difference of an integrated of order one [or I(1) in the terminology of Engle and Granger (1987)] variable implies the existence of some feature and that the test for common cycles in a set of I(1) variables is essentially a test for the existence of common features — features are data properties such as seasonality, heteroscedasticity, autoregressive conditional heteroscedasticity, and serial correlation.

In this chapter, we follow Engle and Kozicki (1993) and consider testing for a common feature of serial correlation. Therefore, the basic idea behind such a serial correlation (co)feature test is to determine whether a serial correlation feature is present in the first differences of a set of cointegrated I(1) variables and then to examine whether there exists a linear combination of the stationary variables that does not have the serial correlation feature. If the linear combination of the stationary variables eliminates the feature, it means that the feature is common across the stationary variables and that they were generated by similar (stationary) stochastic processes. Evidence to the contrary provides strong empirical support that the series are generated by significantly different (stationary) stochastic processes.

Suppose, for example, that in our bivariate setting the y_t and x_t series are I(1) variables and that each series has been rendered stationary by removing the stochastic trend. We can write equations (14.1) and (14.2) as

$$\Delta y_t = c_{yt} + \epsilon_{yt}$$

$$\Delta x_t = c_{xt} + \epsilon_{xt}.$$

Assuming that the cyclical component is common across the two series, $c_{xt} = \beta c_{yt}$ where β is the factor of proportionality between the cyclical components, a linear combination between Δy_t and Δx_t can be written as

$$\Delta z_t = c_{yt} + \epsilon_{yt} - \mu \left(\beta c_{yt} + \epsilon_{xt} \right)$$

$$= c_{yt} - \mu \beta c_{yt} + \epsilon_{yt} - \mu \epsilon_{xt},$$

which for $\mu = 1/\beta$ reduces to a series made up of the noise components. The test for a common serial correlation feature is thus a test of whether there is some 'cofeature vector' $[1, \mu]$ for which Δz_t does not have the serial correlation feature.

14.2.3   Codependent Cycles

In introducing the notion of common features, Engle and Kozicki (1993) expand on the work by Engle and Granger (1987) on common trends and cointegration and provide a test for the existence of common cycles. However, as Ericsson (1993, p. 380) argues, in an early critique of the Engle and Kozicki (1993) methodology, common feature tests have some shortcomings and that

> "... detecting the presence of a cofeature depends on the dating of the series. If the relative lag between the series is not correct, a test for a cofeature may fail to find a cofeature when there is one, even asymptotically."

To illustrate, suppose that the Δy_t and Δx_t series have exactly the same serial correlation cofeature but at different lags, as follows

$$\Delta y_t = c_{yt} + \epsilon_{yt}$$

$$\Delta x_t = \beta c_{yt-k} + \epsilon_{xt}.$$

In this case, a linear combination of Δy_t and Δx_t at time t will not remove the feature even though each of the Δy_t and Δx_t series individually has the same feature. If, however, Δy_t enters the linear combination at lag k, as follows,

$$\Delta z_t = c_{yt-k} + \epsilon_{yt-k} - \mu \left(\beta c_{yt-k} + \epsilon_{xt} \right)$$

$$= c_{yt-k} - \mu\beta c_{yt-k} + \epsilon_{yt-k} - \mu\epsilon_{xt},$$

then for $\mu = 1/\beta$ the serial correlation common feature is eliminated from the Δz_t series. Vahid and Engle (1997) refer to the presence of a lagged serial correlation cofeature of this kind as a 'codependent cycle.'

A codependent cycle is not as strong a form of comovement as a common cycle. It provides, however, a stronger and more informative test of underlying comovements between a group of variables than traditional (lagged) cross-correlation analysis does. In what follows, we test for common trends, common cycles, and (where appropriate) codependent cycles in North American natural gas and crude oil markets.

14.3 The Evidence

We use daily data from January 1991 to April 2001 on spot Henry Hub natural gas and WTI crude oil prices — see Figure 14.1 for a graphical presentation of the series. The first step in examining trends between crude oil and natural gas prices is to test for the presence of a stochastic trend (a unit root) in the autoregressive representation of each individual series. In doing so, we use two alternative unit root testing procedures to deal with anomalies that arise when the data are not very informative about whether or not there is a unit root.

In the first two columns of Table 14.1 we report p-values for the augmented Dickey-Fuller (ADF) tests [see Dickey and Fuller (1981)] and the nonparametric, $Z(t_{\hat{\alpha}})$, test of Phillips (1987) and Phillips and Perron (1988). These p-values (calculated using TSP 4.5) are based on the response surface estimates given by MacKinnon (1994). For the ADF test, the optimal lag length was taken to be the order selected by the Akaike information criterion (AIC) plus 2; see Pantula *et al.* (1994) for details regarding the advantages of this rule for choosing the number of augmenting lags. The $Z(t_{\hat{\alpha}})$ test is done with the same Dickey-Fuller regression variables, using no augmenting lags.

TABLE 14.1
MARGINAL SIGNIFICANCE LEVELS OF
ADF AND $Z(\mathrm{T}_{\hat{\alpha}})$ UNIT ROOT TESTS

Series	Log levels		Logged differences	
	ADF	$Z(t_{\hat{\alpha}})$	ADF	$Z(t_{\hat{\alpha}})$
WTI oil	.517	.310	.000	.000
Henry Hub gas	.074	.019	.000	.000

Notes: Sample period, daily data: 01/01/1991-26/04/2001.
Numbers are tail areas of tests.

Based on the p-values for the ADF and $Z(t_{\hat{\alpha}})$ test statistics reported in Table 14.1, the null hypothesis of a unit root in log levels cannot be rejected. However, the null hypothesis of a unit root in the first logged differences is rejected, in the last two columns of Table 14.1, suggesting that the series are difference stationary. This is consistent with the Nelson and Plosser (1982) argument that most macroeconomic and financial time series have a stochastic trend.

Next we use the Pesaran *et al.* (2001) autoregressive distributed lag, bounds test approach to the problem of testing for the existence of a long-run relationship between Henry Hub natural gas and WTI crude oil prices.[1] As already noted, this approach has the advantage of testing for long-run relations without requiring that the underlying variables are stationary or integrated. To briefly describe the methodology, consider a vector error correction model

$$\Delta \mathbf{Y}_t = \mu + \psi t + \lambda \mathbf{Y}_{t-1} + \sum_{j=1}^{p-1} \gamma_j \Delta \mathbf{Y}_{t-j} + \varepsilon_t, \qquad (14.5)$$

where $\mathbf{Y}_t = [y_t \ x_t]'$, where (as before) y_t is the logged natural gas price and x_t the logged crude oil price. $\mu = [\mu_y \ \mu_x]'$ is a vector of constant terms, $\Delta = 1 - L$, and

---

[1]Coe and Serletis (2001) have also used the Pesaran *et al.* (2001) methodology in the context of absolute and relative purchasing power parity tests.

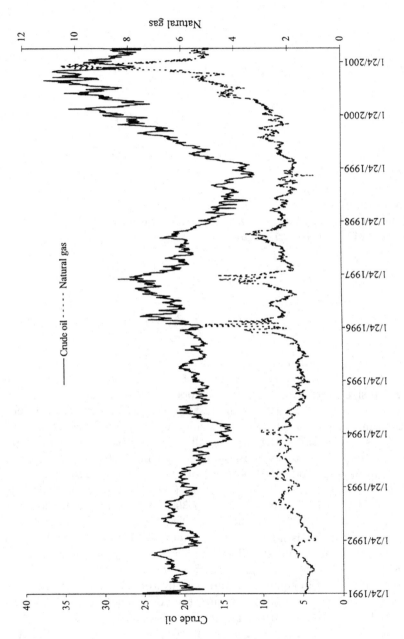

Figure 14.1: WT1 oil and Henry Hub natural gas prices, 1991-2001.

$$\gamma_j = \left[\begin{array}{cc} \gamma_{yy,j} & \gamma_{yx,j} \\ \gamma_{xy,j} & \gamma_{xx,j} \end{array} \right] = -\sum_{k=j+1}^{p} \phi_k.$$

Here λ is the long-run multiplier matrix and is given by

$$\lambda = \left[\begin{array}{cc} \lambda_{yy} & \lambda_{yx} \\ \lambda_{xy} & \lambda_{xx} \end{array} \right] = -\left(\mathbf{I} - \sum_{j=1}^{p} \phi_j \right),$$

where $\mathbf{I}$ is a 2×2 identity matrix. The diagonal elements of this matrix are left unrestricted. This allows for the possibility that the series can be either I(0) or I(1) — for example, $\lambda_{yy} = 0$ implies that y_t is I(1) and $\lambda_{yy} < 0$ implies that it is I(0).

Under the assumption that $\lambda_{xy} = 0$, the equation for natural gas from equation (14.5) can be written as

$$\Delta y_t = \alpha_0 + \alpha_1 t + \varphi y_{t-1} + \delta x_{t-1} + \omega \Delta x_t$$

$$+ \sum_{j=1}^{p-1} \beta_{yj} \Delta y_{t-j} + \sum_{j=1}^{q-1} \beta_{xj} \Delta x_{t-j} + u_t, \qquad (14.6)$$

where $\alpha_0 = \mu_y - \omega \mu_x$, $\alpha_1 = \psi_y - \omega \psi_x$, $\varphi = \lambda_{yy}$, $\delta = \lambda_{yx} - \omega \lambda_{xx}$, $\beta_{yj} = \gamma_{yy,j} - \omega \gamma_{xy,j}$ and $\beta_{xj} = \gamma_{yx,j} - \omega \gamma_{xx,j}$. This can also be interpreted as an autoregressive distributed lag (ARDL) model. We estimate equation (14.6) by ordinary least squares (OLS) and test the absence of a long-run (levels) relationship between y_t and x_t, by calculating the F statistic for the null hypothesis of $\varphi = \delta = 0$ (against the alternative that $\varphi \neq 0$ and $\delta \neq 0$). The distribution of this test statistic under the null depends on the order of integration of y_t and x_t. If both y_t and x_t are I(0), the asymptotic 5% critical value is 6.56 — see Pesaran et al. (2001, Table C1.v). If both y_t and x_t are I(1), the 5% critical value is 7.30. For cases in which one series is I(0) and one is I(1), the critical value falls in the interval [6.56, 7.30].

In practice, there is no reason why p and q in equation (14.6) should have the same value, and we allow for this possibility. In particular, we consider values from 1 to 15 (given the high-frequency nature of the data) for each of p and q in equation (14.6), and by running 225 regressions we choose the specifications that minimize the AIC value. The AIC selects the ARDL (14, 1) specification and the F-statistic for the joint significance of φ and δ is 9.49. Since this F-statistic exceeds the upper bound of the critical value band, we can reject the null hypothesis of no long-run relationship between natural gas and crude oil prices, irrespective of the order of their integration. Under the common trends interpretation [see, for example, Stock and Watson (1988)] this result is not too surprising. The same underlying stochastic growth components presumably affect both markets, implying that the WTI crude oil and Henry Hub natural gas markets are linked together.

Next we test the null hypothesis of a common cycle to see whether the series are driven by a common serial correlation process. Before conducting such a test, however, it is important in the first step to establish that the serial correlation feature is present in both series, as it doesn't make sense to test for commonality if the feature is present in only one of the series. In doing so, we follow Engle and Kozicki (1993) and Vahid and Engle (1993) and conduct the serial correlation test in the context of the following VAR framework, in which the natural gas price is treated as jointly determined with the crude oil price,

$$\Delta y_t = \alpha_1 + \alpha_{11}\Delta y_{t-1} + \alpha_{12}\Delta x_{t-1} + \alpha_{13}\hat{\varepsilon}_{t-1} + \zeta_{yt} \qquad (14.7)$$

$$\Delta x_t = \alpha_2 + \alpha_{21}\Delta y_{t-1} + \alpha_{22}\Delta x_{t-1} + \alpha_{23}\hat{\varepsilon}_{t-1} + \zeta_{xt}, \qquad (14.8)$$

where $\hat{\varepsilon}_{t-1}$ is the lagged equilibrium error from the cointegrating regression. The test for a serial correlation feature is a test of whether lagged price changes are significant (i.e., useful in forecasting future price changes). In the LM version of the test, the LM test statistic for the null hypothesis of 'no serial correlation feature' is computed as the coefficient of determination multiplied by the sample size, $T \times R^2$, and is distributed as a χ^2 with three degrees of freedom. The LM feature test statistic is 33.168 for the Δy_t equation and 16.339 for the Δx_t equation, with the 5% critical value being 7.81. Since the test statistic values are greater than the critical value of 7.81, we conclude that there is evidence of a serial correlation feature in each of the Henry Hub natural gas and WTI crude oil prices.

Since we have identified a serial correlation feature in each of the natural gas and crude oil prices, we follow Engle and Kozicki (1993) and Vahid and Engle (1993) and implement the second step of the common cycles test, by estimating by 2SLS and LIML (which are asymptotically equivalent procedures) the following regression equation

$$\Delta y_t = \phi_0 + \phi_1 \Delta x_t + u_t,$$

taking Δy_{t-1}, Δx_{t-1}, and $\hat{\varepsilon}_{t-1}$ as instruments. It turns out [see Engle and Kozicki (1993) and Vahid and Engle (1993) for more details] that the test statistic for a serial correlation common feature is asymptotically equivalent to the test statistic for the legitimacy of the instruments. In fact, the overidentifying test statistic is the $T \times R^2$ from regressing the error term $\hat{u}_t$ on the instruments

$$\hat{u}_t = \vartheta_0 + \vartheta_1 \Delta y_{t-1} + \vartheta_2 \Delta x_{t-1} + \vartheta_3 \hat{\varepsilon}_{t-1} + \xi_t,$$

with the test statistic being distributed as a χ^2 with two degrees of freedom.

Table 14.2 contains three entries for each of the asymptotically equivalent test procedures. It shows the estimated coefficient $\hat{\phi}_1$ (for different dependent variables), its t-statistic, and the LM tests statistic for legitimacy of the instruments.

TABLE 14.2

COMMON CYCLE TESTS BETWEEN WTI OIL AND HENRY HUB GAS

	IV test Dependent variable		LIML test
	Δy_t	Δx_t	Δx_t
$\hat{\phi}_1$	−.083	−.010	−.032
t-statistic	−.166	−.116	−.272
LM statistic	33.008	16.300	16.264

The LM test statistics exceed the 5% critical value of 5.99, thereby rejecting the null hypothesis of a common cycle.

Although the null hypothesis of a common synchronized cycle has been rejected, it is possible that the Henry Hub natural gas and WTI crude oil markets may face the same cycle but at different speeds, perhaps because of different adjustment costs or different institutional arrangements. We consider therefore whether codependent cycles can be identified for the Henry Hub natural gas and WTI crude oil prices by performing the codependent cycles test discussed in Section 14.2. In doing so, we perform the second step in the test for common features allowing Δy_t to lead and lag Δx_t by up to 48 business days (which is roughly one month).

The results of the codependent cycles test are summarized in Figure 14.2 which plots the values of the LM test statistic (based on LIML estimation) against lags and leads of Henry Hub natural gas price changes. At the 5% level we generally reject the null hypothesis of a codependent cycle for the WTI crude oil and Henry Hub natural gas markets. Thus, to the extent that we reject the null hypotheses of common and codependent cycles, we conclude that deregulation and the increased role of market forces have weakened the relationship between U.S. crude oil and natural gas prices.

14.4 Common Features in Natural Gas Markets

To this point we have omitted any discussion of Canada or Mexico, implicitly assuming that energy prices are the same throughout North America. This is a safe assumption in the case of crude oil, since the WTI crude oil price at Chicago is a North American oil price (or even a world oil price). North American natural gas markets, however, are not as integrated as the oil markets are, because natural gas resources are often located far from demand centers, natural gas is more difficult to transport, and transportation costs are major price-setting concerns.

To provide some evidence on the inter-connectedness of North American natural gas markets, in this section we explore the strength of shared trends and shared cycles between the U.S. Henry Hub and AECO Alberta natural gas markets, using daily data from January 2, 1996 to April 26, 2001 (a total of 1332 observations). The exclusion of Mexico from our analysis and the chosen sample period are based strictly on data availability. Of course, the short length (in calendar time) of these series should be kept in mind in interpreting the results.

Figure 14.3 shows the plots of U.S. Henry Hub and AECO Alberta natural gas prices over the January 1996 to April 2001 period, and Table 14.3 reports unit root test results in the same fashion as those in Table 14.1. Clearly, the Henry Hub natural gas price series has a unit root but the AECO series is not very informative about its unit root properties; the ADF test cannot reject the unit root null at the 5% level whereas the $Z(t_{\hat{\alpha}})$ test rejects it. Also, using the Pesaran *et al.* (2001) bounds testing approach, we reject the null hypothesis of a lack of a long-run relationship between Henry Hub and AECO natural gas prices — the F-statistic is 10.05.

To explore the strength of common features between the U.S. Henry Hub and AECO Alberta natural markets, we first test for a serial correlation feature in each of the Henry Hub and AECO natural gas prices, in the context of the VAR framework of equations (14.7) and (14.8). The LM feature test statistic is 11.003 for the Δ(Henry Hub price) equation and 57.984 for the Δ(AECO price) equation, with the 5% critical value being 7.81. Thus, we conclude that there is evidence of a serial correlation feature in each of the Henry Hub and AECO natural gas prices.

TABLE 14.3
MARGINAL SIGNIFICANCE LEVELS OF
ADF AND $Z(t_{\hat{\alpha}})$ UNIT ROOT TESTS

Series	Log levels		Logged differences	
	ADF	$Z(t_{\hat{\alpha}})$	ADF	$Z(t_{\hat{\alpha}})$
Henry Hub gas	.410	.166	.000	.000
AECO gas	.076	.002	.000	.000

Notes: Sample period, daily data: 02/01/1996-26/04/2001.
Numbers are tail areas of tests.

Finally, in Table 14.4 we test for a common cycle between the U.S. Henry Hub and AECO Alberta natural markets. The results are reported in the same fashion as those in Table 14.2. We find that we cannot reject the null hypothesis of a common synchronized cycle, suggesting that North Americn gas markets are driven by a common serial correlation feature. This is consistnt with the observation (in Figure 14.3) that there is a strong correlation between the U.S. Henry Hub and AECO Alberta natural gas prices. Moreover, it suggests that Henry Hub natural gas prices could be characterized as North American natural gas prices, in the same way that the WTI crud oil prices at Chicago are characterized as North American crude oil prices.

TABLE 14.4
COMMON CYCLE TESTS BETWEEN HENRY HUB AND AECO GAS

	IV test		LIML test
	Dependent variable		
	Δ (AECO gas)	Δ (HH gas)	Δ (HH gas)
$\hat{\phi}_1$	−2.474	−.254	−.279
t-statistic	−2.646	−2.319	2.438
LM statistic	4.106	3.153	3.106

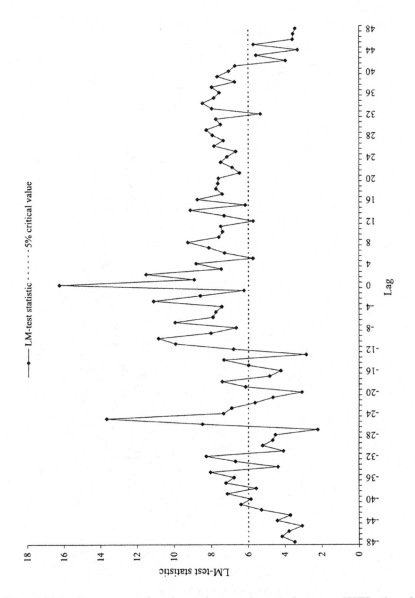

Figure 14.2: Common and codependent cycle tests between WTI oil and Henry Hub natural gas.

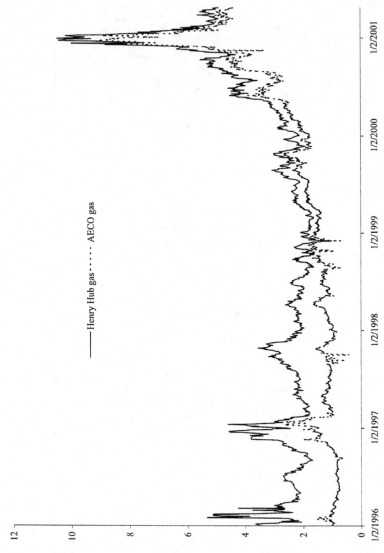

Figure 14.3: U.S. Henry Hub and AECO Alberta natural gas prices, 1996-2001.

14.5   Conclusion

We have investigated the strength of shared trends and shared cycles between WTI crude oil prices and Henry Hub natural gas prices using daily data from January, 1990 to April, 2001. Based on the testing procedures recently suggested by Engle and Kozicki (1993) and Vahid and Engle (1993), we have rejected the null hypotheses of common and codependent cycles, suggesting that there has been 'de-coupling' of the prices of these two energy sources as a result of oil and gas deregulation in the United States.

We also tested for a common cycle between the U.S. Henry Hub and AECO Alberta natural gas markets, in an attempt to investigate the interconnectedness of North American natural gas markets. We could not reject the null hypothesis of a common synchronized cycle, implying a high degree of similarity in the impulse responses of U.S. Henry Hub and AECO Alberta natural gas prices to cycle generating innovations. This result also confirms the hypothesis that in the deregulated period North American natural gas prices are largely defined by the U.S. Henry Hub price trends.

Part 5

Volatility Modelling
in Energy Markets

Overview of Part 5

Apostolos Serletis

The following table contains a brief summary of the contents of the chapters in Part 5 of the book. This part of the book consists of two chapters that use recent advances in the financial econometrics literature.

Volatility Modelling in Energy Markets

Chapter Number	Chapter Title	Contents
15	Returns and Volatility in the NYMEX Henry Hub Natural Gas Futures Market	This chapter provides a study of the determinants of daily returns and volatility in the NYMEX natural gas market, using recent advances in the financial econometrics literature.
16	Measuring and Testing Natural Gas and Electricity Markets Volatility: Evidence from Alberta's Deregulated Markets	Chapter 16 builds on recent contributions by Grier *et al.* (2004) and Shields *et al.* (2005) and specifies and estimates a multivariate GARCH-M model of natural gas and electricity price changes, and tests for causal relationships between natural gas and electricity price changes and their volatilities.

Chapter 15:

This chapter uses autoregressive conditional heteroscedasticity (ARCH)-type models to investigate the determinants of returns and volatility in the NYMEX Henry Hub natural gas futures contract market. Using daily data, for the period that natural gas has been traded on the exchange, it finds significant evidence of seasonal and open interest effects in both returns and volatility.

Chapter 16:

This chapter specifies and estimates a multivariate GARCH-M model of natural gas and electricity price changes, and tests for causal relationships between natural gas and electricity price changes and their volatilities, using data over the deregulated period from January 1, 1996 to November 9, 2004 from Alberta's (deregulated) spot power and natural gas markets. The model allows for the possibilities of spillovers and asymmetries in the variance-covariance structure for natural gas and electricity price changes, and also for the separate examination of the linear and nonlinear effects of changes in natural gas and electricity prices.

Chapter 15

Returns and Volatility in the NYMEX Henry Hub Natural Gas Futures Market

*Apostolos Serletis and Asghar Shahmoradi**

15.1 Introduction

Recently economists have been creating new models and tools that can capture important nonlinearities in economic and financial data. There have been, for example, exciting advances in dynamical systems theory, nonlinear time-series analysis, and stochastic volatility models. One reason for the interest in nonlinear methods is what one might call the 'forecasting paradox' — the fact that linear models produce invariably good in-sample fits, but usually fail miserably at out-of-sample prediction. One is therefore tempted to explore means by which apparent dependencies in the residuals of linear models (that are inconsistent with a linear data generator) can be exploited to produce better forecasts.

In this chapter we use recent advances in the financial econometrics literature and conduct a thorough investigation to properly identify the type of heteroscedasticity in the data generation process of natural gas

*Originally published in *OPEC Review* (2006), 171–186. Reprinted with permission.

futures prices. In particular, we use Engle's (1982) autoregressive conditional heteroscedasticity (ARCH) model and Bollerslev's (1986) extension to a generalized ARCH (GARCH) model to model time varying returns and volatility in the NYMEX Henry Hub natural gas futures contract market. Moreover, we follow contributions by Milonas (1986, 1991), Gay and Kim (1987), Malick and Ward (1987), Kenyon *et al.* (1987), and Liew and Brooks (1998) and investigate the determinants of daily returns and volatility in this market.

Several recent studies have applied models from the ARCH/GARCH family of models in modeling time varying volatility in high frequency financial data — for example, Bollerslev, Chou, and Kroner (1992) probably list well over 100 papers employing ARCH-type techniques to fit financial time series. Although there have been some applications in the context of energy prices, as for example in Deaves and Krinsky (1992) and Day and Lewis (1993, 1997), there is no study that explicitly studies any aspects of the NYMEX Henry Hub natural gas futures market. In this chapter, we characterize NYMEX Henry Hub natural gas futures prices as North American natural gas futures prices and test for seasonal, volume, and open interest effects in daily returns and volatility over the period that natural gas has been traded on NYMEX.

The remainder of the chapter is organized as follows. The next section describes the data and presents some descriptive statistics for the returns series. Sections 15.3 and 15.4 provide the necessary theoretical background and model the returns and volatility of NYMEX natural gas futures prices, by specifying parametric ARCH/GARCH-type models for volatility. Section 15.5 summarizes the chapter.

15.2   The Data

We use daily NYMEX Henry Hub natural gas futures contract data, from Norman's Historical Data (http://www.normanshistoricaldata.com), and construct a continuous series of one month natural gas futures prices by using the rollover approach at the delivery date of the nearest to maturity futures contract. We do not model the one month natural gas futures price, z_t, directly but instead we model returns by taking the logarithmic first difference of that price, $\Delta \log z_t$. We use daily data from April 30, 1990 to June 27, 2002 — a total of 2755 observations.

In Tables 15.1, 15.2, and 15.3 we report summary statistics for daily, monthly, and annual returns, respectively. The descriptive statistics in Table 15.1 show a day of the week effect, with returns being positive only on Friday, perhaps due to the release of weekly storage information. In fact, the lowest return is observed on Tuesday and the highest on Friday. The monthly results in Table 15.2 are broadly consistent with the winter cycle in natural gas that runs from November through March. In these months, where there is the greatest uncertainty about supply and demand, average returns are negative and the variance of returns is relatively high.

The behavior of annual returns (see Table 15.3) reflects primarily developments in North American natural gas markets, given that the bulk of natural gas is consumed in North America. It is to be noted that natural gas

TABLE 15.1

DESCRIPTIVE STATISTICS FOR DAILY RETURNS

	Mean	Variance
Mon	-.00034	.00042
Tue	-.00087	.00024
Wed	-.00052	.00023
Thu	-.00008	.00020
Fri	.00114	.00020

TABLE 15.2

DESCRIPTIVE STATISTICS FOR MONTHLY RETURNS

	Mean	Variance
Jan	-.00253	.00045
Feb	-.00067	.00028
Mar	.00199	.00017
Apr	.00067	.00013
May	-.00003	.00010
Jun	-.00001	.00017
Jul	-.00081	.00016
Aug	.00063	.00019
Sep	.00253	.00037
Oct	.00186	.00028
Nov	-.00057	.00023
Dec	-.00189	.00064

TABLE 15.3
DESCRIPTIVE STATISTICS FOR YEARLY RETURNS

	Mean	Variance
1990	.00074	.00019
1991	-.00080	.00013
1992	.00058	.00022
1993	.00027	.00019
1994	-.00023	.00021
1995	.00084	.00025
1996	.00019	.00060
1997	-.00042	.00024
1998	-.00027	.00022
1999	.00032	.00017
2000	.00255	.00023
2001	-.00253	.00043
2002	.00108	.00026

markets are more segmented than crude oil markets in the sense that when North American crude oil prices change, they tend to change world-wide, whereas the price of natural gas can easily change in North America without any change in natural gas prices on other continents. This follows because transportation of natural gas by pipeline is cheaper than transportation by ship (liquefied natural gas).

15.3   Modeling Returns

Having determined the presence of seasonal effects in returns, we use the following general autoregressive (AR) model to model the mean of returns

$$\Delta \log z_t = \varphi_0 + \sum_{i=1}^{r} \varphi_i \Delta \log z_{t-i}$$

$$+ \sum_{j=1} d_j D_{jt} + \sum_{j=1} m_j M_{jt} + \sum_{j=1} y_j Y_{jt}$$

$$+ \delta_1 \text{CVOL}_t + \delta_2 \text{MVOL}_t + \delta_3 \text{OPIN}_t + \varepsilon_t, \qquad (15.1)$$

where D_{jt} are day of the week dummy variables, M_{jt} are month of the year dummy variables, and Y_{jt} are yearly dummy variables. CVOL_t is the volume of the nearest to maturity futures contract, MVOL_t is the market volume (that is, the volume in all traded contracts), and OPIN_t is the open interest of all contracts traded at the different maturities. r is the order of the autoregression, $(\varphi, d, m, y, \delta)$ are unknown parameters to be estimated, and ε_t is a random shock, assumed to be $\mathrm{IN}(0, \sigma_\varepsilon^2)$. The optimal lag length of the autoregression, r, was selected using the Akaike Information Criterion (AIC) and was set equal to 3 — that is, $r = 3$ in (15.1).

The results of estimating equation (15.1) using ordinary least squares (OLS) are presented in Table 15.4. We see that the lagged returns are statistically significant and that there are significant seasonal effects. In particular, at conventional significance levels, there are day of the week effects, month of the year effects (in that five months, March, April, August, September, and October are statistically significant), and year effects in 1990, 1991, 1992, 1993, 1995, and 2000. Moreover, there are significant open interest effects, but the volume effects are found to be statistically insignificant.

Regarding the year effects in 1990, 1991, 1992, 1993, 1995, and 2000, those from 1990 to 1995 potentially reflect the reduced investments in exploration and production in the early 1990s compared with investments in the 1980s and the effects of the Persian Gulf war. Due in part to mild weather, there was no growth in gas consumption from 1996 and 1999. The year effect in 2000 is consistent with higher wellhead prices, California's environmental regulations on electricity generators that added to gas demand, and to higher overall natural gas demand — in fact, that demand was met by a large net drawdown of gas in storage and an increase in imports.

In order to achieve a more parsimonious model for the mean of returns, we test the joint significance of various effects by performing F tests of whether a subset of the included variables in (15.1) all have zero coefficients. The results of these tests are presented in Table 15.5. They are generally consistent with the individual parameter results and indicate that in a joint test only the day of the week and volume effects are not statistically significant. As a result, in what follows we use equation (15.1) with $d_j = \delta_1 = \delta_2 = 0$ as a more parsimonious model for the mean of returns; the results of estimating this model are presented in Table 15.6. Notice that the fit of equation (15.1) in Tables 15.4 and 15.6 is bad, as indicated by the R^2. This bad fit is to be expected, however, since $\Delta \log z_t$ is a 'noisy' time series.

TABLE 15.4

OLS PARAMETER ESTIMATES OF EQUATION (15.1)

Variable	Coefficient	t-statistic
φ_0	-.00884	-2.144
$\Delta \log z_{t-1}$	-.04213	-2.181
$\Delta \log z_{t-2}$	-.05760	-2.992
$\Delta \log z_{t-3}$	-.03963	-2.059
Mon	-.00195	-1.958
Tue	-.00163	-1.676
Wed	-.00213	-2.194
Thu	-.00183	-1.857
Jan	-.00028	-.180
Feb	.00205	1.272
Mar	.00519	3.283
Apr	.00354	2.259
May	.00240	1.578
Jun	.00240	1.583
Jul	.00172	1.113
Aug	.00296	1.958
Sep	.00465	3.005
Oct	.00449	2.950
Nov	.00136	.859
1990	.00853	2.079
1991	.00668	1.723
1992	.00756	2.054
1993	.00593	1.775
1994	.00449	1.435
1995	.00508	1.727
1996	.00384	1.364
1997	.00199	.773
1998	.00120	.518
1999	.00031	.162
2000	.00353	1.742
2001	-.00181	-.860
CVOL	5.55E-08	.501
MVOL	4.03E-09	.047
OPIN	1.05E-07	2.853

$R^2 = 0.023$, $DW = 1.995$.

TABLE 15.5
F-Tests of Various Combinations of
Parameter Estimates in Eq. (15.1)

Variables	F-statistic	p-value
$\Delta \log z_{t-i}$	2.209	.085
Day effect	1.784	.129
Month effect	2.065	.019
Year effect	1.848	.036
CVOL, MVOL	.940	.390

15.4 Modeling Volatility

So far, we have assumed that the natural gas price series has a constant variance (that is, it is homoscedastic, as opposed to heteroscedastic) and determined a model for the mean of returns. Many macroeconomic and financial variables, however, exhibit clusters of volatility and tranguility (i.e., serial dependence in the higher conditional moments), and in such circumstances the homoscedasticity assumption is inappropriate.

To illustrate the unsatisfactory nature of standard econometric models for modeling risk and uncertainty, consider the following first order autoregressive model

$$y_t = \phi_0 + \phi_1 y_{t-1} + \varepsilon_t, \quad \varepsilon_t \sim N(0, \sigma^2),$$

assuming that $|\phi_1| < 1$ for stationarity, and suppose that we want to forecast y_{t+1}. The unconditional forecast of y_{t+1} (always being the long-run mean of the sequence) is simply $\phi_0/(1 - \phi_1)$ and the unconditional forecast error variance (i.e., the long-run forecast of the variance) is $\sigma^2/(1 - \phi_1)$.

Instead, if conditional forecasts are used, the conditional forecast of y_{t+1} is $\phi_0 + \phi_1 y_t$ and the conditional forecast error variance is σ^2. Clearly, the unconditional and conditional forecast error variances are different, unless $\phi_1 = 0$, but they are both constants — they do not depend on the available information set and hence do not change over time. In fact, since $1/(1 - \phi_1) > 1$, the unconditional forecast has a greater variance than the conditional forecast, meaning that conditional forecasts are preferable (since they take into account the known current and past realizations of series).

TABLE 15.6
RESTRICTED OLS PARAMETER ESTIMATES OF EQUATION (15.1)

Variable	Coefficient	t-statistic
φ_0	-.00566	-1.849
$\Delta \log z_{t-1}$	-.03890	-1.575
$\Delta \log z_{t-2}$	-.05587	-2.135
$\Delta \log z_{t-3}$	-.03876	-1.498
Jan	-.00062	-.279
Feb	.00168	.806
Mar	.00473	2.423
Apr	.00313	1.654
May	.00216	1.182
Jun	.00215	1.139
Jul	.00145	.765
Aug	.00279	1.454
Sep	.00465	2.156
Oct	.00442	2.149
Nov	.00148	.734
1990	.00401	1.479
1991	.00229	.934
1992	.00340	1.391
1993	.00235	1.037
1994	.00128	.589
1995	.00223	1.047
1996	.00125	.524
1997	-.00011	-.058
1998	-.00033	-.182
1999	-.00056	-.328
2000	.00235	1.310
2001	-.00329	-1.626
OPIN	8.14E-08	2.497

$R^2 = 0.020$, $DW = 1.998$.

Since the vast improvement in forecasts due to time series models, stems from the use of the conditional mean, one might expect better forecasts with a model in which the unconditional variance is constant but the conditional variance, like the conditional mean, is also a random variable depending on current and past information. A model which allows the conditional variance to depend on the past realization of the series is the autoregressive conditional heteroscedasticity (ARCH) model introduced by Engle (1982), according to which the conditional variance is assumed to depend on lagged values of squared residuals, as follows

$$\sigma_t^2 = w_0 + \sum_{i=1}^{p} \alpha_i \varepsilon_{t-i}^2, \qquad (15.2)$$

with $p \geq 0$ (for $p = 0$, ε_t is simply white noise) and u_t and $\varepsilon_{t-i}, i = 1, ..., p$ independent. Note that the disturbances in the ARCH(p) model are serially uncorrelated but not independent, as they are related through second moments.

An extension of the ARCH model is the generalized ARCH, or GARCH, model proposed by Bollerslev (1986). In the generalized ARCH (p, q) model — called GARCH(p, q) — we have

$$\sigma_t^2 = w_0 + \sum_{i=1}^{p} \alpha_i \varepsilon_{t-i}^2 + \sum_{j=1}^{q} \beta_j \sigma_{t-j}^2, \qquad (15.3)$$

where $w_0 > 0$, $\alpha_i \geq 0$, $i = 1, ..., p$, and $\beta_j \geq 0$, $j = 1, ..., q$. In (16.2) the conditional variance is assumed to depend on lagged values of squared residuals and also on lagged values of itself — an autoregressive component is introduced.

Having selected an optimal model for the mean of returns, equation (15.1) with $d_j = \delta_1 = \delta_2 = 0$, we now proceed to formally test the residuals of that model for the presence of an ARCH-type process, before we can use the class of ARCH/GARCH models to model volatility. We do so, using Engle's (1982) Lagrange multiplier test for ARCH-type disturbances. This involves regressing the squared residuals from the autoregression (15.1), with $d_j = \delta_1 = \delta_2 = 0$, against a constant and q lagged values of the squared residuals, as follows

$$\hat{\varepsilon}_t^2 = w_0 + \sum_{i=1}^{p} \alpha_i \hat{\varepsilon}_{t-i}^2 + u_t \qquad (15.4)$$

If there are no ARCH-type effects, the estimated coefficients α_1 through α_p would be equal to zero, meaning that this regression will have little explanatory power and the coefficient of determination, R^2, will be very low. If the sample size is T, under the null hypothesis of no ARCH-type errors, the test statistic $T \times R^2$ converges to a χ_p^2 distribution. If $T \times R^2$ is sufficiently large, rejection of the null hypothesis that the coefficients of the lagged squared residuals are all equal to zero is equivalent to rejecting the null hypothesis of no ARCH-type errors. In fact, as shown by Lee (1991) this test is also the Lagrange multiplier test for GARCH-type disturbances, where the null hypothesis is $\alpha_1 = \cdots = \alpha_p = \beta_1 = \cdots = \beta_q = 0$.

Using 1, 2, 5 and 10 lags in equation (15.4), the Lagrange multiplier test rejects the null hypothesis that the coefficients of the lagged squared residuals are all equal to zero, suggesting the existence of an ARCH/GARCH-type process in the residuals.

Of course, any number of ARCH or GARCH models are likely to be suitable for modeling these effects. To optimally select a particular model from the ARCH/GARCH family of models, we proceed as follows. We use maximum likelihood estimation techniques to estimate ARCH models ranging from ARCH (1) to ARCH (15) and GARCH models ranging from GARCH (1,1) to GARCH (3,3). We rule out those models where the parameter estimates fail to converge as well as those models where a particular parameter failed to estimate because of singularity problems. Finally, we apply the AIC to the remaining models in order to choose the preferred model. Following these steps, we choose an ARCH(4) as the preferred ARCH model and a GARCH (2,1) as the preferred GARCH model.

Next, we use conditional volatility estimates, $\hat{h}_t$, generated from each of the ARCH(4) and GARCH(2,1) models to estimate the following equation, using ordinary least squares,

$$\hat{h}_t = \varphi_0 + \sum_{j=1} d_j D_{jt} + \sum_{j=1} m_j M_{jt} + \sum_{j=1} y_j Y_{jt}$$

$$+ \delta_1 \text{CVOL}_t + \delta_2 \text{MVOL}_t + \delta_3 \text{OPIN}_t + \varepsilon_t, \qquad (15.5)$$

where $v_t \sim \text{IN}(0, \sigma_\varepsilon^2)$ and the other variables are defined as in equation (15.1). We present the results of estimating (15.5) in Table 15.7 and report F-tests of the joint significance of the various effects in Table 15.8, in the same fashion as we did in Table 15.5 for equation (15.2). We find that all the effects are statistically significant at conventional significance levels, except for the volume effects in the ARCH(4) model and the daily and volume effects in the GARCH(2,1) model.

15.5   Conclusion

This chapter provides a study of the determinants of daily returns and volatility in the NYMEX natural gas market over the period from April 30, 1990 to June 5, 2002, using recent advances in the financial econometrics literature. The contribution of the chapter is its use of models of changing volatility to properly identify the type of heteroscedasticity in the data-generation processes. Our results strongly support the presence of seasonal

TABLE 15.7

OLS PARAMETER ESTIMATES OF EQUATION (15.5)

Variable	ARCH(4)		GARCH(2,1)	
	Coefficient	t-statistic	Coefficient	t-statistic
φ_0	.00033	10.720	.00057	13.107
Mon	-1.15E-05	-1.542	-1.12E-05	-1.061
Tue	9.74E-06	1.332	9.69E-06	.938
Wed	-4.35E-06	-.595	6.31E-06	.611
Thu	-1.66E-06	-.223	8.11E-06	.774
Jan	3.51E-05	2.931	-1.91E-06	-.113
Feb	-.00012	-10.606	-.00024	-14.356
Mar	-.00019	-16.396	-.00032	-19.141
Apr	-.00022	-19.107	-.00032	-19.756
May	-.00027	-23.841	-.00037	-23.294
Jun	-.00019	-17.117	-.00031	-19.370
Jul	-.00018	-15.910	-.00029	-17.853
Aug	-.00017	-15.788	-.00029	-18.241
Sep	-3.63E-05	-3.116	-.00011	-6.873
Oct	-2.65E-05	-2.317	-.00012	-7.480
Nov	-.00012	-10.400	-.00024	-14.428
1990	-1.11E-06	-.035	-3.41E-05	-.785
1991	-6.34E-05	-2.180	-.00020	-4.967
1992	2.37E-05	.858	-.00010	-2.770
1993	1.90E-05	.759	-.00011	-3.168
1994	-1.32E-05	-.563	-.00012	-3.750
1995	-3.50E-05	-1.587	-.00015	-5.068
1996	.00012	5.989	.00014	4.882
1997	1.20E-05	.621	-9.74E-05	-3.563
1998	-3.08E-05	-1.757	-.00018	-7.262
1999	-7.70E-05	-5.303	-.00021	-10.513
2000	-1.88E-05	-1.235	-.00016	-7.433
2001	.00010	6.463	1.43E-05	.639
CVOL	-4.02E-10	-.483	-6.93E-10	-.590
MVOL	5.58E-10	.877	6.36E-10	.709
OPIN	8.00E-10	2.883	1.01E-09	2.572

$R^2 = .467, DW = 1.06.$ $R^2 = .466, DW = .542.$

TABLE 15.8

F-TESTS OF VARIOUS COMBINATIONS OF
PARAMETER ESTIMATES IN EQ. (15.5)

Variables	ARCH(4)		GARCH(2,1)	
	F-statistc	p-value	F-statistc	p-value
Day effect	2.158	.071	1.285	.273
Month effect	147.57	.000	126.186	.000
Year effect	50.311	.000	74.937	.000
CVOL, MVOL	1.021	.360	.269	.750

and open interest effects in both returns and volatility, consistent with evidence of previous research on other futures markets — see, for example, Najand and Yung (1991), Foster (1995), and Liew and Brooks (1998).

Although some critics of futures markets suggest that the low cost of trading in these markets induces excessive speculation, causing higher market volatility, we believe that the large capital requirements and significant lead times associated with the production and delivery of energy make these markets very sensitive to the imbalances between demand and supply capability, thereby resulting in price volatility. Moreover, weather conditons and capacity constraints are affecting the natural gas market in such a way that we observe high market volatility — see, for example, Serletis (1997) and Serletis and Shahmoradi (2005, 2006).

We have carried out univariate analysis on NYMEX natural gas prices without exploiting the covariance between natural gas and other energy markets. This suggests that a particularly constructive approach would potentially be based on the use of a higher dimensional system that exploits the covariance among different energy markets such as crude oil markets, electricity markets, coal markets, and perhaps renewable markets. Focusing on higher dimensional ARCH/GARCH-type modeling in the context of energy markets and updating our analysis to capture the recent high volatility in the natural gas market is an area for potentially productive future research.

Chapter 16

Measuring and Testing Natural Gas and Electricity Markets Volatility: Evidence from Alberta's Deregulated Markets

*Apostolos Serletis and Akbar Shahmoradi**

16.1   Introduction

Recent leading-edge research has applied various innovative methods for modeling spot wholesale electricity prices — see, for example, Deng and Jiang (2004), León and Rubia (2004), Serletis and Andreadis (2004), and Hinich and Serletis (2006). These works are interesting and attractive, but have taken a univariate time series approach to the analysis of electricity prices. From an economic perspective, however, the interest in the price of electricity is in its relationship with the prices of various underlying primary fuel commodities. As Bunn (2004, p. 2) recently put it

*Originally published in *Studies in Nonlinear Dynamics and Econometrics* 10(3) (2006), Article 10. Reprinted with permission.

" ··· take the case of gas, for example. This is now becoming the fuel of choice for electricity generation. The investment costs are lower than coal, or oil plant; it is cleaner and, depending upon location, the fuel costs are comparable. But with more and more of the gas resources being used for power generation, in some markets the issue of whether gas drives power prices, or *vice versa*, is not easily answered."

Investigating the relationship between electricity and natural gas prices (and their volatilities) is our primary objective in this chapter. Since natural gas is an input in electricity generation, it is expected natural gas price changes to be (at least partly) reflected in electricity price changes. The same argument applies to the relationship between natural gas price uncertainty and electricity price uncertainty. Therefore, the investigation of the behavior of electricity prices requires that we take into account the behavior of natural gas prices, which rules out the possibility of relying on a single equation approach. Moreover, to investigate the effects of uncertainty on realizations of natural gas and electricity prices, we jointly model the conditional variance-covariance process underlying natural gas and electricity price changes.

In doing so, we build on recent contributions by Grier *et al.* (2004) and Shields *et al.* (2005) and specify and estimate a multivariate GARCH-M model of natural gas and electricity price changes, and test for causal relationships between natural gas and electricity price changes and their volatilities, using data over the deregulated period from January 1, 1996 to November 9, 2004 from Alberta's (deregulated) spot power and natural gas markets. The model allows for the possibilities of spillovers and asymmetries in the variance-covariance structure for natural gas and electricity price changes, and also for the separate examination of the linear and nonlinear effects of changes in natural gas and electricity prices.

The chapter is organized as follows. Section 16.2 describes the data and Section 16.3 provides a description of the multivariate GARCH-M model that we use to test for causality between natural gas and electricity price changes and their volatilities. Section 16.4 presents and discusses the empirical results. The final section briefly concludes the chapter.

16.2   The Data

We use hourly electricity prices (sourced from the Alberta Power Pool), denominated in megawatt-hours (MWh) and concentrate on Alberta's peak power market (in order to capture the relationship between natural gas and power), which is a 6 day per week and 16 hours per day market — Monday through Saturday from 8:00 a.m. to 11:00 p.m. Because the Alberta natural gas data is only available for weekdays and non-holidays, we aggregated the power data for weekdays and non-holidays only. For natural gas, AECO is the most liquid intra-provincial index and daily spot prices were obtained from Bloomberg. The sample period is from January 1, 1996 to November 9, 2004.

Table 16.1 presents summary statistics for the levels and changes of natural gas and electricity prices and Figures 16.1 and 16.2 plot electricity and natural gas prices, respectively. As can be seen in the first panel of Table 16.1, electricity prices are more volatile than natural gas prices, and there is significant evidence of skewness and excess kurtosis in both series and their changes, with all series failing to satisfy the null hypothesis of the Bera-Jarque (1980) test for normality. The lower panel of Table 16.1 presents Ljung–Box (1979) tests for serial correlation indicating that there is a significant amount of serial dependence in both levels and changes of natural gas and electricity prices. Similarly a Ljung–Box test for serial correlation in the squared data provides strong evidence of conditional heteroscedasticity in the data.

Finally, Engle and Granger (1987) cointegration tests (not reported here) suggest that the null hypothesis of no cointegration between electricity and natural gas prices is rejected at conventional significance levels, suggesting an error correction representation between these series.

16.3 The Model

We use a general asymmetric GARCH-in Mean model of natural gas and electricity price changes that allows for the possibilities of spillovers and asymmetries in the variance-covariance structure of natural gas and electricity prices. In particular, we use an extended version of a VARMA (vector autoregressive moving average) GARCH in mean model, in natural gas price changes (g_t) and electricity price changes(e_t), as follows

$$\boldsymbol{y}_t = \boldsymbol{a} + \boldsymbol{b}\,\hat{\varepsilon}_{t-1} + \sum_{i=1}^{p}\boldsymbol{\Gamma}_i\boldsymbol{y}_{t-i}$$

$$+ \sum_{j=0}^{q}\boldsymbol{\Psi}_j\boldsymbol{h}_{t-j} + \sum_{k=1}^{r}\boldsymbol{\Phi}_k\boldsymbol{z}_{t-k} + \sum_{l=1}^{s}\boldsymbol{\Theta}_l\boldsymbol{u}_{t-l} + \boldsymbol{u}_t \qquad (16.1)$$

TABLE 16.1

SUMMARY STATISTICS OF DAILY NATURAL GAS AND
ELECTRICITY PRICES

Series	Stand. Dev.	Skewness	Excess Kurtosis	Jarque-Bera
P_g	2.277	1.041 [0.000]	1.476 [0.000]	599.10 [0.000]
P_e	79.354	3.271 [0.000]	12.721 [0.000]	18810. [0.000]
ΔP_g	0.074	1.243 [.000]	41.319 [.000]	157500 [.000]
ΔP_e	0.441	-0.023 [.000]	3.824 [.000]	1344.59 [.000]

Ljung-Box tests of unconditional correlations

	$Q(4)$	$Q^2(4)$	$Q(12)$	$Q^2(12)$
P_g	8504.11 [.000]	7812.02 [.000]	24411.1 [.000]	21126.1 [.000]
P_e	4202.79 [.000]	2929.17 [.000]	10585.3 [.000]	7038.91 [.000]
ΔP_g	158.63 [.000]	215.17 [.000]	175.25 [.000]	221.58 [.000]
ΔP_e	281.02 [.000]	431.22 [.000]	294.26 [.000]	628.43 [.000]

Note: Numbers in parentheses are p-values. $Q(4)$ and $Q^2(4)$ are Q-statistics for testing serial correlation in the residuals and the squared residuals, respectively.

with

$$\boldsymbol{u}_t \,|\, \Omega_{t-1} \sim (\boldsymbol{0}, \boldsymbol{H}_t), \qquad \boldsymbol{H}_t = \begin{bmatrix} h_{g_t} & h_{ge_t} \\ h_{ge_t} & h_{e_t} \end{bmatrix},$$

where Ω_{t-1} denotes the available information set in period $t-1$ and

$$\boldsymbol{y}_t = \begin{bmatrix} g_t \\ e_t \end{bmatrix}; \boldsymbol{u}_t = \begin{bmatrix} u_{g_t} \\ u_{e_t} \end{bmatrix}; \boldsymbol{h}_t = \begin{bmatrix} h_{g_t} \\ h_{e_t} \end{bmatrix}; \boldsymbol{a} = \begin{bmatrix} a_g \\ a_e \end{bmatrix};$$

$$\boldsymbol{\Gamma}_i = \begin{bmatrix} \gamma_{11}^{(i)} & \gamma_{12}^{(i)} \\ \gamma_{21}^{(i)} & \gamma_{22}^{(i)} \end{bmatrix}; \boldsymbol{\Psi}_j = \begin{bmatrix} \psi_{11}^{(j)} & \psi_{12}^{(j)} \\ \psi_{21}^{(j)} & \psi_{22}^{(j)} \end{bmatrix};$$

$$\boldsymbol{\Phi}_k = \begin{bmatrix} \phi_{11}^{(k)} & \phi_{12}^{(k)} \\ \phi_{21}^{(k)} & \phi_{22}^{(k)} \end{bmatrix}; \boldsymbol{\Theta}_l = \begin{bmatrix} \theta_{11}^{(l)} & \theta_{12}^{(l)} \\ \theta_{21}^{(l)} & \theta_{22}^{(l)} \end{bmatrix};$$

$$\boldsymbol{z}_{t-k} = \begin{bmatrix} z_{g_{t-k}} \\ z_{e_{t-k}} \end{bmatrix}; z_{j_{t-k}} = \frac{u_{j_{t-k}}}{\sqrt{h_{j_{t-k}}}}, \text{ for } j = g, e.$$

Notice that h_{t-j} and z_{t-k} have been introduced to take anticipated and unanticipated volatilities into account and $\hat{\varepsilon}_{t-1}$ is the error correction term from the long run cointegrating regression.

As in Grier *et al.* (2004) and Shields *et al.* (2005), we introduce an asymmetry into the conditional variance-covariance process in order to deal with good and bad news about natural gas and electricity price changes. In particular, if natural gas price changes are higher than expected, we take that to be bad news. We therefore capture bad news about natural gas price changes by a positive natural gas price change residual, by defining $\epsilon_{g_t} = \max\{u_{g_t}, 0\}$. We also capture bad news about electricity price changes by defining $\epsilon_{e_t} = \max\{u_{e_t}, 0\}$.

It is to be noted that because natural gas is an input in electricity generation, $\epsilon_{g_t} = \max\{u_{g_t}, 0\}$ might be bad news for electricity producers and then partly for consumers, but it should be good news for natural gas producers. In this regard, we also estimated the model for the case where $\epsilon_{g_t} = \min\{u_{g_t}, 0\}$, and we got the same test results, although the estimated coefficients were quantitatively and sometimes qualitatively different.

Following Grier *et al.* (2004), we allow for asymmetric responses as follows

$$H_t = C'C + \sum_{j=1}^{f} B_j' H_{t-j} B_j$$

$$+ \sum_{k=1}^{\kappa} A_k' u_{t-k} u_{t-k}' A_k + D' \epsilon_{t-1} \epsilon_{t-1}' D \qquad (16.2)$$

where C, B_j, A_k, and D are $n \times n$ matrices (for all values of j and k), with C being a triangular matrix to ensure positive definiteness of H. There are $n^2(p+q+r+s+1) + n(n+1)/2 + n^2(f+\kappa+1)$ parameters in (16.1)-(16.2) and in order to deal with estimation problems in the large parameter space we assume that $f = \kappa = 1$ in equation (16.2), consistent with recent empirical evidence regarding the superiority of GARCH(1,1) models — see, for example, Hansen and Lunde (2005).

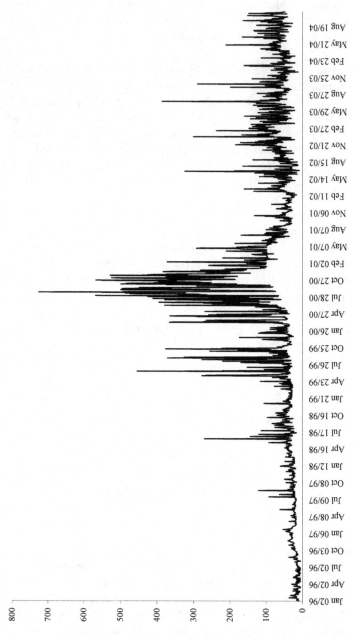

Figure 16.1: Alberta Electricity Prices

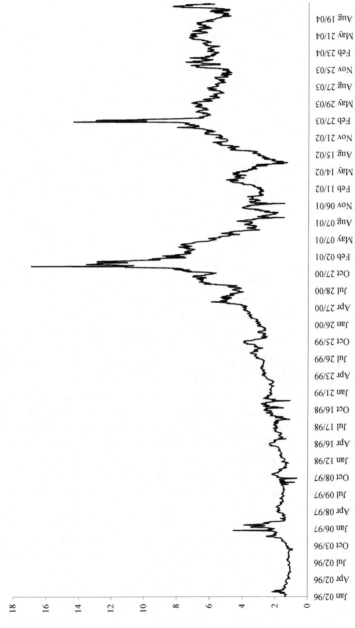

Figure 16.2: AECO Natural Gas Prices

16.4   Empirical Results

We select the optimal values of p, q, r, and s in (16.1) in such a way that there is no serial correlation and ARCH effects in the standardized residuals of the model. In doing so, we choose $p = 4, s = 3, r = 2$, and $q = 1$ in equation (16.1) and $f = \kappa = 1$ in equation (16.2). The inclusion of $\boldsymbol{h}_t$ in (16.1) is consistent with the predictions of finance theory that an asset with a higher perceived risk should pay a higher return on average. The term $\boldsymbol{h}_{t-1}$ is included in (16.1), because it helps, among other lags, to eliminate serial correlation and ARCH effects in the standardized residuals.

In Table 16.2 we report quasi-maximum likelihood (QML) estimates of the parameters (with p-values in parantheses) and diagnostic test statistics, based on the standardized residuals,

$$\hat{z}_{j_t} = \frac{u_{j_t}}{\sqrt{\hat{h}_{j_t}}}, \qquad \text{for } j = e, g.$$

As shown in Table 16.2, the Ljung-Box (1979) Q-statistic for testing serial correlation cannot reject the null of no autocorrelation (at conventional significance levels) for the values and the squared values of the standardized residuals, suggesting that there is no evidence of conditional heteroscedasticity. In addition, the failure of the data to reject the null hypotheses of $E(z) = 0$ and $E(z^2) = 1$, implicitly indicates that the multivariate asymmetric GARCH-M model does not bear significant misspecification error — see, for example, Kroner and Ng (1998).

Figures 16.3-16.5 show the conditional variances for natural gas and electricity price changes as well as the conditional covariance, implied by the estimates of the model. The estimated conditional standard deviations, being the one-period ahead forecasts conditional on past information, are more likely to be a correct representation of future uncertainty than unconditional standard deviations. As can be seen in Figures 16.3 and 16.4, the conditional variance of the electricity price seems to be higher on average than that of the natural gas price. Moreover, for natural gas, volatility appears highest (on average) in 1997 whereas for electricity the period of greatest volatility appears between 1999 and 2001 — a period of increased demand, no excess capacity, and considerable uncertainty about future prices.

Next we examine the model's ability in dealing with potential biases resulting from good and bad news in the natural gas and electricity markets. In doing so, we rely on diagnostic test statistics based on the 'generalized residuals' of Kroner and Ng (1998), defined as $\epsilon_{ij_t} = u_{i_t} u_{j_t} - h_{ij_t}$ for $i, j = e, g$. If our model is specified correctly, it should be able to capture

TABLE 16.2

THE MULTIVARIATE ASYMMETRIC GARCH-M MODEL

Model: Equations (1) and (2) with $p = 4$, $s = 3$, $r = 2$, $q = 1$ and $f = \kappa = 1$

Conditional mean equation

$$a = \begin{bmatrix} -.0009 \\ (.0002) \\ .003 \\ (.002) \end{bmatrix} ; b = \begin{bmatrix} -.0003 \\ (.0001) \\ -.0008 \\ (.0005) \end{bmatrix} ; \Gamma_1 = \begin{bmatrix} -.103 & -.041 \\ (.043) & (.001) \\ .255 & -.950 \\ (.143) & (.006) \end{bmatrix} ; \Gamma_2 = \begin{bmatrix} .115 & -.023 \\ (.076) & (.004) \\ .308 & -.073 \\ (.038) & (.007) \end{bmatrix}$$

$$\Gamma_3 = \begin{bmatrix} .262 & .027 \\ (.027) & (.001) \\ .197 & .409 \\ (.080) & (.016) \end{bmatrix} ; \Gamma_4 = \begin{bmatrix} .029 & .004 \\ (.019) & (.001) \\ .054 & .057 \\ (.029) & (.010) \end{bmatrix} ; \Psi_0 = \begin{bmatrix} 1.610 & -.117 \\ (.169) & (.014) \\ -.049 & -.346 \\ (.179) & (.025) \end{bmatrix}$$

$$\Psi_1 = \begin{bmatrix} -1.365 & .125 \\ (.115) & (.015) \\ -.136 & .343 \\ (.219) & (.025) \end{bmatrix} ; \Phi_1 = \begin{bmatrix} .042 & -.009 \\ (.000) & (.006) \\ .003 & .253 \\ (.006) & (.007) \end{bmatrix} ; \Phi_2 = \begin{bmatrix} -.037 & .006 \\ (.000) & (.000) \\ -.015 & -.199 \\ (.005) & (.007) \end{bmatrix}$$

$$\Theta_1 = \begin{bmatrix} .002 & .053 \\ (.063) & (.000) \\ .141 & .315 \\ (.144) & (.018) \end{bmatrix} ; \Theta_2 = \begin{bmatrix} -.254 & -.0001 \\ (.083) & (.003) \\ .117 & -.604 \\ (.083) & (.011) \end{bmatrix} ; \Theta_3 = \begin{bmatrix} -.256 & -.050 \\ (.037) & (.005) \\ .129 & -.704 \\ (.099) & (.010) \end{bmatrix}$$

Residual diagnostics

	Mean	Variance	$Q(4)$	$Q^2(4)$	$Q(12)$	$Q^2(12)$
u_{gt}	-.009 [.651]	1.005 [.998]	12.71 [.012]	1.090 [.895]	20.71 [.064]	4.329 [.976]
u_{et}	-.010 [.633]	.992 [.997]	1.526 [.821]	5.629 [.228]	4.351 [.976]	9.501 [.659]

Conditional variance-covariance structure

$$C = \begin{bmatrix} .009 & .006 \\ (.000) & (.005) \\ & .028 \\ & (.001) \end{bmatrix} ; B = \begin{bmatrix} .860 & -.012 \\ (.002) & (.008) \\ .0003 & .975 \\ (.0004) & (.008) \end{bmatrix} ;$$

$$A = \begin{bmatrix} .524 & .061 \\ (.022) & (.032) \\ -.007 & .150 \\ (.001) & (.032) \end{bmatrix} ; D = \begin{bmatrix} .197 & .088 \\ (.058) & (.030) \\ -.006 & -.195 \\ (.003) & (.006) \end{bmatrix}$$

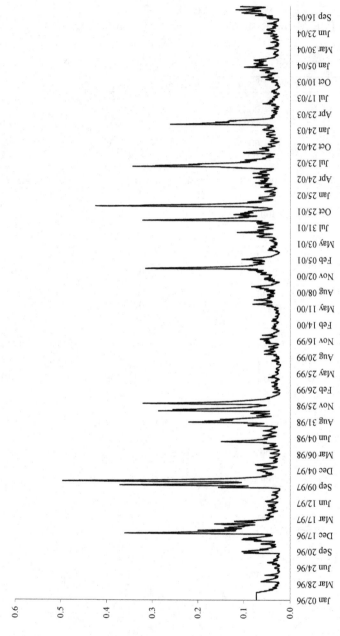

Figure 16.3: Volatility of Natural Gas Price Changes (Conditional
Standard Deviation)

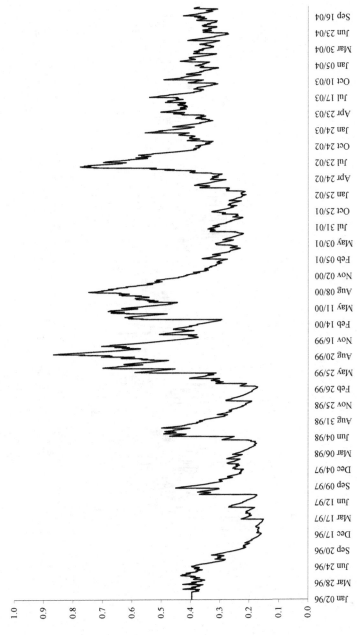

Figure 16.4: Volatility of Electricity Price Changes (Conditional Standard Deviation)

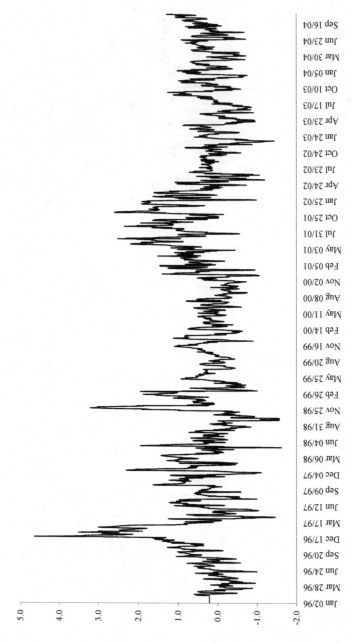

Figure 16.5: Conditional Correlation Between Natural Gas and
Electricity Price Changes

the effect of any kind of good and bad news in the natural gas and electricity markets, resulting in no sign bias. In fact, in all symmetric GARCH models, the news impact curve — see Engle and Ng (1993) — is symmetric and centered at $\epsilon_{i_{t-1}} = 0$. A generalized residual can be thought of as the distance between a point on the scatter plot of $u_{i_t} u_{j_t}$ from a corresponding point on the news impact curve. If the conditional heteroscedasticity part of the model is correct, then $E_{t-1}(u_{i_t} u_{j_t} - h_{ij_t}) = 0$ for all values of i and j. In other words, h_{ij_t} is the conditional expectation of $u_{i_t} u_{j_t}$. For example, if the model (16.1)-(16.2) gives a covariance news impact surface — a three dimensional graph of h_{eg_t} against u_{e_t} and u_{g_t} — which is too low whenever the shock to the natural gas price changes is negative ($u_{g_t} < 0$), then the vertical distance between h_{eg_t} and $u_{e_t} u_{g_t}$ will tend to be positive.

The Engle and Ng (1993) and Kroner and Ng (1998) misspecification indicators test whether we can predict the generalized residuals by some variables observed in the past, but which are not included in the model — this is exactly the intuition behind $E_{t-1}(u_{i_t} u_{j_t} - h_{ij_t}) = 0$. In this regard, we follow Kroner and Ng (1998) and Shields *et al.* (2005) and define two sets of misspecification indicators. In a two dimensional space, we first partition $(u_{e_{t-1}}, u_{g_{t-1}})$ into four quadrants in terms of the possible sign of two residuals. Then to shed light on *any* possible *sign bias* of the model, we define the first set of indicator functions as $I(u_{e_{t-1}} < 0)$, $I(u_{g_{t-1}} < 0)$, $I(u_{e_{t-1}} < 0; u_{g_{t-1}} < 0)$, $I(u_{e_{t-1}} > 0; u_{g_{t-1}} < 0)$, $I(u_{e_{t-1}} < 0; u_{g_{t-1}} > 0)$ and $I(u_{e_{t-1}} > 0; u_{g_{t-1}} > 0)$, where $I(\cdot)$ equals one if the argument is true, and zero otherwise. Significance of any of these indicator functions indicates that the model (16.1)-(16.2) is incapable of predicting the effects of some shocks to either e_t or g_t. Moreover, due to the fact that the possible effect of a shock could be a function of both the *size* and the *sign* of the shock, we define $u_{e_{t-1}}^2 I(u_{e_{t-1}} < 0)$, $u_{e_{t-1}}^2 I(u_{g_{t-1}} < 0)$, $u_{g_{t-1}}^2 I(u_{e_{t-1}} < 0)$, and $u_{g_{t-1}}^2 I(u_{g_{t-1}} < 0)$. These indicators are technically scaled versions of the former ones, with the magnitude of the shocks as a scale measure.

We conducted 33 misspecification indicator tests. As can be seen in Table 16.3, all indicators (except those in 7 cases) fail to reject (at the 5% level) the null of no misspecification — all test statistics in Table 16.3 are distributed as $\chi^2(1)$. Hence, our model (16.1)-(16.2) captures the effects of all sign bias and sign-size scale depended shocks in predicting volatility and there is (in general) no significant misspecification error. This means that the exclusion of some other variable, in either y_t or h_t, is not expected to lead to significant misspecification problems.

As can be seen in Table 16.4, the diagonality restriction, $\gamma_{12}^{(i)} = \gamma_{21}^{(i)} = \theta_{12}^{(l)} = \theta_{21}^{(l)} = 0$ for all i, l is rejected, meaning that the data provide strong evidence of the existence of dynamic interactions between e_t and g_t. The null hypothesis of homoscedastic disturbances requires the A, B, and D matrices to be jointly insignificant (that is, $\alpha_{ij} = \beta_{ij} = \delta_{ij} = 0$ for all i, j) and is rejected at the 1% level or better, suggesting that there is significant conditional heteroscedasticity in the data. The null hypothesis of symmetric conditional variance-covariances, which requires all elements of the D matrix to be jointly insignificant (that is, $\delta_{ij} = 0$ for all i, j), is rejected at the 1% level or better, implying the existence of some asymmetries in the data which the model is capable of capturing. Also, the null hypothesis of a diagonal covariance process requires the off-diagonal elements of the A, B, and D matrices to be jointly insignificant (that is, $\alpha_{12} = \alpha_{21} = \beta_{12} = \beta_{21} = \delta_{12} = \delta_{21} = 0$), but these estimated coefficients are jointly significant at the 1% level or better.

TABLE 16.3

DIAGNOSTIC TESTS BASED ON THE NEWS IMPACT CURVE

	$u_{e_t} u_{e_t} - h_{e_t}$	$u_{e_t} u_{g_t} - h_{eg_t}$	$u_{g_t} u_{g_t} - h_{g_t}$
$E(\cdot)$	.129	.922	.445
$I(u_{e_{t-1}} < 0)$	.245	.040	.067
$I(u_{g_{t-1}} < 0)$	.680	.399	.031
$I(u_{e_{t-1}} < 0; u_{g_{t-1}} < 0)$	.000	.024	.752
$I(u_{e_{t-1}} > 0; u_{g_{t-1}} < 0)$	.067	.690	.033
$I(u_{e_{t-1}} < 0; u_{g_{t-1}} > 0)$	.233	.401	.791
$I(u_{e_{t-1}} > 0; u_{g_{t-1}} > 0)$	.844	.247	.382
$u_{e_{t-1}}^2 I(u_{e_{t-1}} < 0)$	.682	.278	.323
$u_{e_{t-1}}^2 I(u_{g_{t-1}} < 0)$	.387	.870	.000
$u_{g_{t-1}}^2 I(u_{e_{t-1}} < 0)$	.171	.185	.246
$u_{g_{t-1}}^2 I(u_{g_{t-1}} < 0)$	.403	.022	.151

Note: Numbers are tail areas of tests.

Finally, in order to establish the causal relationship between electricity and natural gas prices, in the second half of Table 16.4 (under Granger causality tests) we test a number of null hypotheses. In particular, we test the null hypothesis that electricity price changes do not linearly cause natural gas price changes, $\gamma_{12}^{(i)} = \theta_{12}^{(l)} = 0$ for $i = 1, 2, 3$ and $l = 1, 2$, the null hypothesis that electricity price changes do not nonlinearly cause natural gas price changes, $\psi_{12}^{(1)} = \psi_{12}^{(2)} = \phi_{12}^{(1)} = \phi_{12}^{(2)} = \alpha_{12} = \beta_{12} = \delta_{12} = 0$, and finally the joint null that electricity price changes do not cause natural gas

TABLE 16.4

HYPOTHESES TESTING AND GRANGER CAUSALITY TESTS

Model: Equations (1) and (2) with $p=4$, $s=3$, $r=2$, $q=1$ and $f=\kappa=1$

Hypotheses testing

Diagonal VARMA	$H_0: \gamma_{12}^{(i)} = \gamma_{21}^{(i)} = \theta_{12}^{(l)} = \theta_{21}^{(l)} = 0$, for all i and l	.000
No GARCH	$H_0: \alpha_{ij} = \beta_{ij} = \delta_{ij} = 0$, for all i,j	.000
No GARCH-M	$H_0: \psi_{ij}^k = \phi_{ij}^k = 0$, for all i,j,k	.000
No asymmetry	$H_0: \delta_{ij} = 0$, for $i,j = 1,2$	.000
Diagonal GARCH	$H_0: \alpha_{12} = \alpha_{21} = \beta_{21} = \beta_{12} = \delta_{12} = \delta_{21} = 0$	.000

Granger causality tests

Causality from electricity to natural gas

No linear causality:	$H_0^1: \gamma_{12}^{(i)} = \theta_{12}^{(l)} = 0$, for all i and l	.000
No nonlinear causality:	$H_0^2: \psi_{12}^{(j)} = \phi_{12}^{(k)} = \alpha_{12} = \beta_{12} = \delta_{12} = 0$, for all j and k	.000
No causality:	$H_0: H_0^1 + H_0^2$	.000

Causality from natural gas to electricity

No linear causality:	$H_0^1: \gamma_{21}^{(i)} = \theta_{21}^{(l)} = 0$, for all i and l	.000
No nonlinear causality:	$H_0^2: \psi_{21}^{(j)} = \phi_{21}^{(k)} = \alpha_{21} = \beta_{21} = \delta_{21} = 0$, for all j and k	.000
No causality:	$H_0^3: H_0^1 + H_0^2$	.000

Note: Numbers (in the last column) are tail areas of tests.

price changes (which is the joint application of both the linear and nonlinear restrictions). The roles of electricity and natural gas are reversed in another set of tests to see whether there is a feedback relationship among these variables. All the causality tests are carried out in terms of the Lagrange multiplier principle.

The results in Table 16.4 indicate that there is bidirectional (linear and nonlinear) causality between natural gas and electricity price changes.

16.5 Conclusions

This chapter provides a study of the relationship between natural gas and electricity price changes and their volatilities, using recent advances in the financial econometrics literature. In the context of a VARMA GARCH-in mean model, we jointly model the conditional variance-covariance process underlying natural gas and electricity price changes. Our model provides a good statistical description of the conditional mean and conditional variance-covariance processes characterizing natural gas and electricity price changes.

The model indicates that there is bidirectional (linear and nonlinear) causality between natural gas and electricity prices. It is to be noted that we interpet causality in terms of predictability and not as implying an underlying structural economic relationship between natural gas and electricity prices and their volatilities. Thus, the existence of bidirectional causality between natural gas and electricity prices means that there are empirically effective arbitraging mechanisms in Alberta's natural gas and power markets, raising questions about the efficient markets hypothesis.

This chapter has allowed study of the joint determination of electricity and natural gas prices and focused on relationships between uncertainty about natural gas and electricity prices and their average outcomes. It rules out alternative volatility models that dot not allow for the possibilities of spillovers and asymmetries in the variance-covariance matrix for natural gas and electricity price changes. This is important in volatility measurement — one of the most important issues in the whole of finance — with significant implications for policy and risk management.

Part 6

Chaos, Fractals, and Random Modulations in Energy Markets

Overview of Part 6

Apostolos Serletis

The following table contains a brief summary of the contents of the chapters in Part 6 of the book. Part 6 of the book consists of three chapters that apply tests from statistics and dynamical systems theory to examine the behaviour of energy prices.

Chaos, Fractals, and Random Modulations in Energy Markets

Chapter Number	Chapter Title	Contents
17	The North American Natural Gas Liquids Markets are Chaotic	It tests for deterministic chaos in seven Mont Belview, Texas hydrocarbon markets, using the Lyapunov exponent estimator of Nychka *et al.* (1992).
18	Random Fractal Structures in North American Energy Markets	It uses various tests from statistics and dynamical systems theory to support a random fractal structure for North American energy markets.
19	Randomly Modulated Periodic Signals in Alberta's Electricity Market	This chapter uses hourly electricity prices and MW hour demand for Alberta to test for randomly modulated periodicity. It detects relatively steady weekly and daily cycles in demand, but very unstable cycles in prices.

Chapter 17:

This chapter tests for deterministic chaos (i.e., nonlinear deterministic processes which look random) in seven Mont Belview, Texas hydrocarbon markets, using monthly data drom 1985:1 to 1996:12 — the markets are those of ethane, propane, normal butane, iso-butane, naptha, crudel oil, and natural gas. In doing so, it uses the Lyapunov exponent estimator of Nychka, Ellner, Gallant, and McCaffrey (1992). It concludes that there is evidence consistent with a chaotic nonlinear generation process in all five natural gas liquids markets.

Chapter 18:

This chapter uses daily observations on West Texas Intermediate (WTI) crude oil prices at Chicago and Henry Hub natural gas prices at Louisiana (over the deregulated period of the 1990s) and various tests from statistics and dynamical systems theory to support a random fractal structure for North American energy markets. In particular, this evidence is supported by the Vassilicos *et al.* (1994) multifractal structure test and the Ghashghaie *et al.* (1996) turbulent behavior test.

Chapter 19:

This last chapter uses hourly electricity prices and MW hour demand for Alberta, Canada over the deregulated period after 1996 to test for randomly modulated periodicity. In doing so, it applies the signal coherence spectral analysis to the time series of hourly spot prices and megawatt-hours (MWh) demand from 1/1/1996 to 12/7/2003 using the FORTRAN 95 program developed by Hinich (2000). It detects relatively steady weekly and daily cycles in demand but very unstable cycles in prices.

Chapter 17

The North American Natural Gas Liquids Markets are Chaotic

*Apostolos Serletis and Periklis Gogas**

17.1 Introduction

In the recent years, interest in deterministic chaos (i.e., nonlinear deterministic processes which look random) has increased tremendously and the literature is still growing. Besides its obvious intellectual appeal, chaos represents a radical change of perspective in the explanation of fluctuations observed in economic and financial time series. In this view, the fluctuations and irregularities observed in such series receive an endogenous explanation and are traced back to the strong nonlinear deterministic structure that can pervade the economic system. Moreover, if chaos can be shown to exist, the implication would be that (nonlinearity-based) prediction is possible (at least in the short run and provided the actual generating mechanism is known exactly). Prediction, however, over long periods is all but impossible, due to the 'sensitive dependence on initial conditions' property of chaos

Until recently chaotic dynamics had been studied almost exclusively by theoreticians. However, theorizing might be viewed (by economists) as

---

*Originally published in *The Energy Journal* 20 (1999), 83-103. Reprinted with permission.

empty if there is no evidence of chaos in macroeconomic and financial time series.

Therefore, a number of researchers have recently focused on testing for nonlinearity in general and chaos in particular in economic and financial time series, with encouraging results, especially in the case of financial time series. For example, Scheinkman and LeBaron (1989) studied United States weekly returns on the Center for Research in Security Prices (CRSP) value-weighted index, and found rather strong evidence of nonlinearity and some evidence of chaos. Similar results have been obtained by Frank and Stengos (1989), investigating daily prices for gold and silver. More recently, Serletis and Gogas (1997) test for chaos in seven East European black-market exchange rates and find evidence consistent with a chaotic nonlinear generation process in two out of the seven series — the Russian ruble and East German mark. Barnett and Serletis (1999) provide a state-of-the-art review of this literature.

In this chapter we test for deterministic chaos in North American hydro-carbon markets. In doing so, we use monthly data, from 1985:1 to 1996:12, on Mont Belview, Texas ethane (C2), propane (C3), normal butane (nC4), iso-butane (iC4), naptha (C5), crude oil, and natural gas prices. In the last decade, the North American hydrocarbon industry has seen a dra-matic transformation from a highly regulated environment to one which is more market-driven, and this transition has led to the emergence of differ-ent markets (especially for natural gas and natural gas liquids) throughout North America — see Serletis (1997), for example, for more details. How-ever, capacity constraints seem to be distorting these markets raising the possibility of chaotic price behavior, arising from within the structure of these markets.

The chapter is organized along the following lines. Section 17.2 pro-vides some background regarding North American hydrocarbon markets. Section 17.3 discusses some basic data facts and investigates the univariate time series properties of Belview hydrocarbon prices, interpreting the re-sults in terms of the permanent/temporary nature of shocks. Section 17.4 provides a description of the key features of the Nychka et al. (1992) Lya-punov exponent estimator, focusing explicit attention on the test's ability to detect chaos. Section 17.5 presents the results of the chaos tests and the final section concludes with some suggestions for potentially useful future empirical research.

17.2 Background

The raw natural gas that comes from wells consists mainly of methane (C_1). However, it also contains various quantities of other heavier hydrocarbons such as ethane (C_2), propane (C_3), butane (C_4), and pentane plus (C_5^+) — the subscripts correspond to the number of carbon atoms that the respective gas molecule contains. Moreover, butane can take one of two forms (isomers), normal butane (nC_4) and isobutane (iC_4). These heavier products (with respect to methane) are collectively known as natural gas liquids (NGLs), with C_3 and C_4 often referred to as liquified petroleum gases (LPGs).

NGLs are extracted from raw natural gas in mixed streams. For example, a C_2^+ stream contains C_2, C_3, C_4, and C_5 while a C_3^+ stream contains all of the above except C_2. In fact, some liquids extraction from raw natural gas is necessary in order to meet minimum (gas) pipeline quality specifications. Also, the majority of the C_3^+ is removed from raw natural gas to prevent condensation of these liquids in gas pipelines. Of course, the amount of processing depends on how 'wet' or 'dry' the raw gas is — gas that is rich in NGLs is referred to as 'wet,' whereas gas with a lower than average NGL content is referred to as 'dry' or 'lean.'

Liquids production depends on raw natural gas production, which depends on geographic distribution across basins. In the last decade, the North American natural gas industry has seen a dramatic transformation from a highly regulated industry to one which is more market-driven. The transition to a less regulated, more market-oriented environment has led to the emergence of different spot markets throughout North America. In particular, producing area spot markets have emerged in Alberta, British Columbia, Rocky Mountain, Anadarko, San Juan, Permian, South Texas, and Louisiana basins. Moreover, production sites, pipelines and storage services are more accessible today, thereby ensuring that changes in market demand and supply are reflected in prices on spot, futures, and swaps markets.

Liquids markets, however, have their own dynamics. For example, the fuels do not compete at any of the major burnertips and what has been done to restructure the North American natural gas business has little to do with liquids markets. Capacity constraints, however, that distort North American natural gas markets impact production of natural gas and thus processed liquids. For example, the development of spot markets for natural gas and of storage facilities has had an effect on propane markets, especially the use of propane for peaking and enriching of lean gas streams. Also, on the demand side, there is not a large consumer market for liquids in the United States and Canada, in the sense that liquids are not a primary domestic or commercial fuel, like they are in other countries.

Our objective in this study is not to examine how the North American hydrocarbon markets are linked together, but to test for deterministic chaos in North American hydrocarbon markets, using Mont Belview, Texas spot prices. One of the most interesting aspects of Belview prices is that they are 'marker' prices for traders from many countries. For example, liquids traders at Petrobras, Brazil's national oil company, use Belview in all of their trading formulas. Moreover, international trading activity is important in the formation of liquids prices at Belview. Brazil, for example, is a huge importer of liquids from the United States (and elsewhere), and liquids constitute almost 80% of domestic fuel use in Brazil (and about 90% in Mexico), suggesting that liquids prices at Belview have more to do with trading factors overseas than with North America.

In what follows, we turn to a discussion of some basic facts and to an investigation of the univariate time series properties of Belview hydrocarbon prices. In Section 17.4, we consider univariate statistical tests for nonlinearity and chaos that have been recently motivated by the mathematics of deterministic nonlinear dynamical systems.

17.3   Basic Facts and Integration Tests

One interesting feature of Belview hydrocarbon prices is the contemporaneous correlation between these prices. These correlations are reported in Table 17.1 for log levels and in Table 17.2 for first differences of log levels. To determine whether these correlations are statistically significant, Pindyck and Rotemberg (1990) is followed and a likelihood ratio test of the hypotheses that the correlation matrices are equal to the identity matrix is performed. The test statistic is

$$-2\ln\left(|R|^{N/2}\right)$$

where $|R|$ is the determinant of the correlation matrix and N is the number of observations. This test statistic is distributed as χ^2 with $0.5q(q-1)$ degrees of freedom, where q is the number of series.

TABLE 17.1

CONTEMPORANEOUS CORRELATIONS BETWEEN LOGGED PRICES

	C2	C3	nC4	iC4	C5	Crude oil	Natural gas
C2	1.000						
C3	0.767	1.000					
nC4	0.686	0.906	1.000				
iC4	0.588	0.821	0.923	1.000			
C5	0.611	0.766	0.869	0.928	1.000		
Crude oil	0.547	0.701	0.823	0.890	0.956	1.000	
Natural gas	0.431	0.437	0.396	0.278	0.289	0.266	1.000

$$\chi^2(21) = 1353.50$$

Note: Monthly data: 1985:1-1996:12.

The test statistic is 1353.50 with a p-value of 0.000 for the logged hydrocarbon prices in Table 17.1, suggesting that the hypothesis that Belview hydrocarbon prices are uncorrelated in log levels is rejected. Turning now to Table 17.2, we see that the test statistic is 849.57 with a p-value of 0.000 for the first differences of the logged prices. Clearly, the null hypothesis that these prices are uncorrelated in first differences of log levels is also rejected.

The first step in testing for nonlinearity and chaos is to test for the presence of a stochastic trend (a unit root) in the autoregressive representation of each individual series. Nelson and Plosser (1982) argue that most macroeconomic and financial time series have a unit root (a stochastic trend), and describe this property as one of being 'difference stationary' (DS) so that the first difference of a time series is stationary. An alternative 'trend stationary' model (TS) has been found to be less appropriate.

In what follows we test the null hypothesis of a stochastic trend against the trend-stationary alternative by estimating by ordinary least-squares (OLS) the following augmented Dickey-Fuller (ADF) type regression (see Dickey and Fuller, 1981)

$$\Delta \log y_t = a_0 + a_2 t + \gamma \log y_{t-1} + \sum_{j=1}^{k} b_j \Delta \log y_{t-j} + \varepsilon_t \qquad (17.1)$$

TABLE 17.2
CONTEMPORANEOUS CORRELATIONS BETWEEN
DIFFERENCED (LOGGED) PRICES

	C2	C3	nC4	iC4	C5	Crude oil	Natural gas
C2	1.000						
C3	0.785	1.000					
nC4	0.702	0.811	1.000				
iC4	0.617	0.725	0.828	1.000			
C5	0.646	0.708	0.777	0.803	1.000		
Crude oil	0.582	0.621	0.701	0.703	0.862	1.000	
Natural gas	0.222	0.172	0.121	0.011	0.005	0.035	1.000

$$\chi^2(21) = 849.57$$

Note: Monthly data: 1985:2-1996:12.

where Δ is the difference operator. The k extra regressors in (17.1) are added to eliminate possible nuisance parameter dependencies in the limit distributions of the test statistics caused by temporal dependencies in the disturbances. The optimal lag length (that is, k) is taken to be the one selected by the Akaike information criterion (AIC) plus 2 — see Pantula *et al.* (1994) for details regarding the advantages of this rule for choosing the number of augmenting lags in equation (17.1).

Table 17.3 presents the results. The first column of Table 17.3 gives the optimal value of k in equation (17.1), based on the AIC plus 2 rule, for each price series. This identifies k to be 3 for C2, nC4, iC4, and C5, 4 for C3, 5 for crude oil, and 10 for natural gas. The t-statistics for the null hypothesis $\gamma = 0$ in equation (17.1) are given under τ_τ, in Table 17.3. Under the null hypothesis that $\gamma = 0$, the appropriate critical value of τ_τ at the 5% level (with 100 observations) is -3.45 — see Fuller (1976, Table 8.5.2). Hence, the null hypothesis of a unit root cannot be rejected for all series.

Since the null hypothesis of a unit root hasn't been rejected, there is a question concerning the test's power in the presence of the deterministic part of the regression (i.e., $a_0 + a_2 t$). In particular, one problem is that the presence of the additional estimated parameters reduces degrees of freedom and the power of the test — reduced power means that we will conclude that the process contains a unit root when, in fact, none is present. Another problem is that the appropriate statistic for testing $\gamma = 0$ depends on which regressors are included in the model.

TABLE 17.3
UNIT ROOT TEST RESULTS

Series	k	Test statistics τ_τ	$t(a_2)$	ϕ_3	τ_μ	Decision
C2	3	-3.09	1.66	5.48	-2.75	I(1)
C3	4	-2.59	2.43	4.64	-1.67	I(1)
nC4	3	-3.33	1.53	6.42	-3.13*	I(0)
iC4	3	-2.83	1.11	4.83	-2.82	I(1)
C5	3	-3.26	0.95	6.22	-3.32*	I(0)
Crude oil	5	-3.20	0.94	6.03	-3.23*	I(0)
Natural gas	10	-1.80	2.74	4.94	-1.16	I(1)

Note: Monthly data: 1985:1-1996:12. All the series are in logs. An asterisk indicates rejection of the null hypothesis at the 5% significance level. τ_τ is the t-statistic for the null hypothesis $\gamma = 0$ in equation (17.1). Under the null hypothesis, the appropriate critical value of τ_τ at the 5% significance level (with 100 observations) is -3.45 — see Fuller (1976, Table 8.5.2). $t(a_2)$ is the t-statistic for the presence of the time trend (i.e., the null hypothesis $a_2 = 0$) in equation (17.1), given the presence of a unit root. The appropriate 95% critical value for $t(a_2)$, given by Dickey and Fuller (1981), is 2.79. The ϕ_3 statistic tests the joint null $a_2 = \gamma = 0$ in equation (17.1). The 95% critical value, given by Dickey and Fuller (1981) is 6.49. Finally, τ_μ is the t-statistic for the null $\gamma = 0$ in equation (17.2). The appropriate 95% critical value of τ_μ is -2.89 — see Dickey and Fuller (1976, Table 8.5.2).

Although we can never be sure of the actual data-generating process, here we follow the procedure suggested by Doldado et al. (1990) for testing for a unit root when the form of the data-generating process is unknown. In particular, since the null hypothesis of a unit root is not rejected, it is necessary to determine whether too many deterministic regressors are included in equation (17.1). We therefore test for the significance of the trend term in equation (17.1) under the null of a unit root, using the $t(a_2)$ statistic in Table 17.3. Under the null that $a_2 = 0$ given the presence of a unit root, the appropriate critical value of $t(a_2)$ at the 5% significance level is 2.79 — see Dickey and Fuller (1981). Clearly, the null cannot be rejected, suggesting that the trend is not significant. The ϕ_3 statistic which tests the joint null hypothesis $a_2 = \gamma = 0$ reconfirms this result.

This means that we should estimate the model without the trend, i.e., in the following form

$$\Delta \log y_t = a_0 + \gamma y_{t-1} + \sum_{j=1}^{k} b_j \Delta \log y_{t-j} + \varepsilon_t \qquad (17.2)$$

and test for the presence of a unit root using the τ_μ statistic. The results, reported in Table 17.3, indicate that the null hypothesis of a unit root is now rejected for nC4, C5, and crude oil. The remaining series do contain a unit root, based on this unit root testing procedure. Our decision regarding the univariate time series properties of these series is summarized in the last column of Table 17.3.

17.4 Tests for Chaos

Recently, five highly regarded tests for nonlinearity or chaos (against various alternatives) have been introduced — see Barnett *et al.* (1995, 1997) for a detailed discussion. All five of the tests are purported to be useful with noisy data of moderate sample sizes. The tests are the Hinich (1982) bispectrum test, the BDS (Brock, Dechert, Scheinkman, and LeBaron, 1996) test, White's (1989) neural network test, Kaplan's (1994) test, and the Nychka, Ellner, Gallant, and McCaffrey (1992) dominant Lyapunov exponent estimator. Another very promising test [that is, similar in some respects to the Nychka, *et al.* (1992) test] has also been recently proposed by Gencay and Dechert (1992).

It is to be noted, however, that the Hinich bispectrum test, the BDS test, White's test, and Kaplan's test are currently in use for testing nonlinear dependence [whether chaotic (i.e., nonlinear deterministic) or stochastic], which is necessary but not sufficient for chaos. Only the Nychka *et al.* (1992) and the Gencay and Dechert (1992) tests are specifically focused on chaos as the null hypothesis. In what follows, we only apply the Lyapunov exponent estimator of Nychka *et al.* (1992). This is a Jacobian-based method involving the use of a neural net to estimate a map function by nonlinear least squares, and subsequently the use of the estimated map and the data to produce an estimate of the dominant Lyapunov exponent. We first describe this test, following Serletis and Gogas (1997).

We assume that the data $\{x_t\}$ are real-valued and are generated by a nonlinear autoregressive model of the form

$$x_t = f(x_{t-L}, x_{t-2L}, \ldots, x_{t-mL}) + e_t \qquad (17.3)$$

where L is the time-delay parameter, m is the length of the autoregression, and e_t is a sequence of zero mean (and unknown constant variance) independent random variables. A state-space representation of (17.3) can be written as follows

$$\begin{pmatrix} x_t \\ x_{t-L} \\ \vdots \\ x_{t-mL+L} \end{pmatrix} = \begin{pmatrix} f(x_{t-L}, \ldots, x_{t-mL}) \\ x_{t-L} \\ \vdots \\ x_{t-mL+L} \end{pmatrix} + \begin{pmatrix} e_t \\ 0 \\ \vdots \\ 0 \end{pmatrix}$$

or equivalently,

$$X_t = F(X_{t-L}) + E_t \tag{17.4}$$

where

$$X_t = (x_t, x_{t-L}, \ldots, x_{t-mL+L})^T,$$

$$F(X_{t-L}) = f((x_{t-L}, \ldots, x_{t-mL}), x_{t-L}, \ldots, x_{t-mL+L})^T,$$

and $E_t = (e_t, 0, \ldots, 0)^T$.

The definition of the dominant Lyapunov exponent, λ, can be formulated more precisely as follows. Let X_0, $X_0' \in R^m$ denote two 'nearby' initial state vectors. After M iterations of model (17.4) with the same random shock we have (using a truncated Taylor approximation)

$$\|X_M - X_M'\| = \|F^M(X_0) - F^M(X_0')\| \simeq \|(DF^M)_{X_0} (X_0 - X_0')\|$$

where F^M is the Mth iterate of F and $(DF^M)_{X_0}$ is the Jacobian matrix of F evaluated at X_0. By application of the chain rule for differentiation, it is possible to show that

$$\|X_M - X_M'\| \simeq \|T_M (X_0 - X_0')\|$$

where $T_M = J_M J_{M-1} \ldots J_1$ and $J (DF^M)_{X_t}$. Letting $\nu_1(M)$ denote the largest eigenvalue of $T_M^T T_M$ the formal definition of the dominant Lyapunov exponent, λ, is

$$\lambda = \lim_{M \to \infty} \frac{1}{2M} \ln |\nu_1(M)|.$$

In this setting, λ gives the long-term rate of divergence or convergence between trajectories. A positive λ measures exponential divergence of two nearby trajectories [and is often used as a definition of chaos — see, for example, Denecker and Pelikan (1986)], whereas a negative λ measures exponential convergence of two nearby trajectories.

In the next section we use the Nychka *et al.* (1992) Jacobian-based method and the LENNS program [see Ellner *et al.* (1992)] to estimate the dominant Lyapunov exponent. In particular we use a neural network model to estimate f by nonlinear least squares, and use the estimated map $\hat{f}$ and the data $\{x_t\}$ to produce an estimate of the dominant Lyapunov exponent. In doing so, we follow the protocol described in Nychka *et al.* (1992).

The predominant model in statistical research on neural nets is the single (hidden) layer feedforward network with a single output. In the present context it can be written as

$$\hat{f}(X_t, \theta) = \alpha + \sum_{j=1}^{k} \beta_j \psi(\omega_j - \gamma_j^T X_t)$$

where $X \in R^m$ is the input, ψ is a known (hidden) univariate nonlinear 'activation function' [usually the logistic distribution function $\psi(u) = 1/(1 + \exp(-u))$ — see, for example, Nychka *et al.* (1992) and Gencay and Dechert (1992)], $\theta = (\alpha, \beta, \omega, \gamma)$ is the parameter vector, and $\gamma_j = (\gamma_{1j}, \gamma_{2j}, \ldots, \gamma_{mj})^T$. $\beta \in R^k$ represents hidden unit weights and $\omega \in R^k$, $\gamma \in R^{k \times m}$ represent input weights to the hidden units. k is the number of units in the hidden layer in the neural net. Notice that there are $[k(m+2)+1]$ free parameters in this model.

Given a data set of inputs and their associated outputs, the network parameter vector, θ, is fit by nonlinear least squares to formulate accurate map estimates. As appropriate values of L, m, and k, are unknown, LENNS selects the value of the triple (L, m, k) that minimizes the Bayesian Information Criterion (BIC) — see Schwartz (1978). Gallant and White (1992) have shown that we can then use $\hat{J}_t$, the estimate of the Jacobian matrix J_t obtained from the approximate map $\hat{f}$, as a nonparametric estimator of J_t. The estimate of the dominant Lyapunov exponent then is

$$\hat{\lambda} = \frac{1}{2N} \ln |\hat{\nu}_1(N)|$$

where $\hat{\nu}_1(N)$ is the largest eigenvalue of $T_N^T T_N$ and where $\hat{T}_N = \hat{J}_N \hat{J}_{N-1} \ldots \hat{J}_1$.

17.5   Empirical Results

Before conducting nonlinear dynamical analysis the data must be rendered stationary, delinearized (by replacing the stationary data with residuals from an autoregression of the data) and transformed (if necessary). Since a stochastic trend has been confirmed for each of C2, C3, iC4, and natural gas, these series are rendered stationary by taking first differences of logarithms.

In the case of C4, C5, and crude oil we use the logged series, since these are I(0). Also, since we are interested in nonlinear dependence, we remove any linear dependence in the stationary data by fitting the best possible linear model. In particular, we prefilter the stationary series by the following autoregression

$$z_t = b_0 + \sum_{j=1}^{q} b_j z_{t-j} + \varepsilon_t, \quad \varepsilon_t \mid I_{t-1} \sim N(0, w_0) \tag{17.5}$$

using for each series the number of lags, q, for which the Ljung-Box (1978) $Q(36)$ statistic is not significant at the 5% level. This identifies q to be 1 for C2 and nC4, 2 for C3, iC4, C5, and crude oil, and 3 for natural gas — see Table 17.4.

TABLE 17.4

DIAGNOSTICS OF AR MODELS UNDER THE
LJUNG-BOX (1978) Q(36) TEST STATISTIC

$$z_t = b_0 + \sum_{j=1}^{q} b_j z_{t-j} + \varepsilon_t, \quad \varepsilon_t \mid I_{t-1} \sim N(0, w_0)$$

| Series | AR Lag, q | AR Error Term Diagnostics (p-values) | | |
		Q-statistic	ARCH	J-B
C2	1	0.532	0.025	0.000
C3	2	0.054	0.802	0.000
nC4	1	0.095	0.057	0.000
iC4	2	0.124	0.097	0.002
C5	2	0.840	0.030	0.000
Crude oil	2	0.639	0.049	0.000
Natural gas	3	0.098	0.035	0.000

Note: The Q-statistic is distributed as a $\chi^2(36)$ on the null of no autocorrelation. ARCH is Engle's (1982) Autoregressive Condidtional Heteroskedasticity (ARCH) test distributed as a $\chi^2(1)$ on the null of no ARCH. The Jarque-Bera test statistic is distributed as a $\chi^2(2)$ under the null hypothesis of normality.

Although the autocorrelation diagnostics in Table 17.4 indicate that the chosen AR models adequately remove linear dependence in the stationary data, the ARCH test suggests the presence of a time-varying variance (except in the case of C3). Since variance-nonlinearity could be generated by either a (stochastic) ARCH process or a deterministic process, in what follows we follow Serletis and Gogas (1997) and model the conditional variance (or predictable volatility) using Bollerslev's (1986) generalized autoregressive conditional heterskedasticity (GARCH) model and Nelson's (1991) exponential GARCH (EGARCH) model. One important feature of what we are doing, however, is to present the results of a diagnostic test for checking the adequacy of these models and choose among the estimated GARCH and EGARCH models.

The GARCH model is a generalization of the pure ARCH model, originally due to Engle (1982) and is useful in detecting nonlinear patterns in variance while not destroying any signs of deterministic structural shifts in a model — see, for example, Lamoureux and Lastrapes (1990). Using the same AR structure as before we estimate the following GARCH(1,1) model

$$z_t = b_0 + \sum_{j=1}^{q} b_j z_{t-j} + \varepsilon_t, \quad \varepsilon_t \mid I_{t-1} - N(0, \sigma_t^2) \qquad (17.6)$$

$$\sigma_t^2 = w_0 + \alpha_1 \varepsilon_{t-1}^2 + \beta_1 \sigma_{t-1}^2$$

where $N(0, \sigma_t^2)$ represents the normal distribution with mean zero and variance σ_t^2. Parameter estimates and diagnostic tests are given in Table 17.5. First, estimated coefficients of the ARCH term, α_1, and the GARCH term, β_1, are positive and (in general) significant at the 5% level. Also, the Q-test finds no linear dependence and the ARCH test finds no ARCH effects, suggesting that the lag structure of the conditional variance is correctly identified. However, the null hypothesis that $\alpha_1 + \beta_1 = 1$ cannot be rejected, suggesting the presence of integrated variances.

GARCH models assume that the conditional variance in equation (17.6) is a function only of the magnitude of the lagged residuals and not their signs — i.e., only the size, not the sign, of lagged residuals determines conditional variance. This assumption imposes important limitations on GARCH models. For example, these models are not well suited to capture the so-called 'leverage effect.' To meet these objections, we use Nelson's (1991) exponential GARCH(1,1), or EGARCH(1,1), also inspired by Engle's (1982) ARCH model, in which the conditional variance σ_t^2 depends on both the size and the sign of lagged residuals as follows

$$\log \sigma_t^2 = w_0 + \beta \log(\sigma_{t-1}^2) + \alpha \left| \frac{\varepsilon_{t-1}}{\sigma_{t-1}} \right| + \gamma \frac{\varepsilon_{t-1}}{\sigma_{t-1}}.$$

The log transformation ensures that σ_t^2 remains non-negative for all t. Clearly, the impact of the most recent residual is now exponential rather than quadratic.

TABLE 17.5

GARCH (1,1) PARAMETER ESTIMATES AND ERROR TERM DIAGNOSTICS

$$z_t = b_0 + \sum_{j=1}^{q} b_j z_{t-j} + \varepsilon_t, \quad \varepsilon_t \mid I_{t-1} \sim N(0, \sigma_t^2), \quad \sigma_t^2 = w_0 + \alpha_1 \varepsilon_{t-1}^2 + \beta_1 \sigma_{t-1}^2$$

| Series | AR Lag, q | GARCH (1,1) Parameter Estimates | | | GARCH (1,1) Error Term Diagnostics (p-values) | | | | | |
		w_0	α_1	β_1	Q-statistic	$Q(\varepsilon^2)$	ARCH	J-B	Log L	$\alpha_1 + \beta_1 = 1$
C2	1	0.000 (1.3)	0.151 (2.4)	0.759 (7.2)	0.877	0.994	0.985	0.000	283.978	0.234
C3	2	0.000 (1.1)	0.053 (0.6)	0.815 (4.5)	0.142	0.999	0.800	0.000	298.829	0.231
nC4	1	0.001 (1.2)	0.132 (1.2)	0.765 (4.2)	0.098	0.964	0.961	0.000	130.292	0.297
iC4	2	0.000 (0.9)	0.091 (0.9)	0.816 (4.2)	0.104	0.769	0.495	0.000	336.847	0.402
C5	2	0.001 (1.3)	0.098 (1.0)	0.676 (3.3)	0.733	0.998	0.928	0.000	159.809	0.147
Crude oil	2	0.001 (2.3)	0.467 (1.4)	0.449 (2.5)	0.064	0.992	0.588	0.013	177.956	0.680
Natural gas	3	0.000 (0.7)	0.920 (3.0)	0.547 (9.1)	0.000	0.997	0.982	0.000	-4.294	0.081

Notes: Numbers in parentheses next to the GARCH (1,1) parameter estimates are absolute t-ratios. The Q-statistic is distributed as a $\chi^2(36)$ on the null of no autocorrelation. The ARCH statistic is distributed as a $\chi^2(1)$ on the null of no ARCH. The Jarque-Bera test statistic is distributed as a $\chi^2(2)$ under the null hypothesis of normality.

Parameter estimates and diagnostic tests for the EGARCH(1,1) model are presented in Table 17.6.   In general, the log likelihood for the EGARCH(1,1) model is higher than that for the GARCH(1,1) model, suggesting that the EGARCH model is superior to the GARCH model for these series. To investigate this further, and in order to choose between GARCH and EGARCH models, we present in Table 17.7 the results of a diagnostic test suggested by Kearns and Pagan (1993) for checking the adequacy of these models. The test involves the regression of $\hat{\varepsilon}_t^2$ against a constant and the estimated conditional variance $\hat{\sigma}_t^2$. The intercept of such a regression should be zero and the slope coefficient unity.

The insignificant $Q(36)$ statistic in Table 17.7 indicates that each of these models captures much of the persistence in actual volatility and the coefficient of determination indicates how well the estimated conditional variance predicts the actual variance and is used to compare the GARCH and EGARCH models. On the basis of these results, and a comparison between the log likelihood values in Tables 17.6 and 17.7, in what follows we test for chaos using the standardized EGARCH(1,1) residuals — the standardized residuals are defined as $\varepsilon_t/\hat{\sigma}_t$, where ε_t is the residual of the mean equation and $\hat{\sigma}_t^2$ its estimated (time-varying) variance.

We now apply the Nychka *et al.* (1992) Lyapunov exponent test to the standardized residuals. The Bayesian Information Criterion (BIC) point estimates of the dominant Lyapunov exponent for each parameter triple (L, m, k) are displayed in Table 17.8 along with the respective optimized value of the BIC criterion. Clearly, all but two Lyapunov exponent point estimates are positive, supporting the conclusion that all Belview natural gas liquids prices have a chaotic nonlinear generating process.

Of course, the standard errors of the estimated dominant Lyapunov exponents are not known [there has not yet been any published research on the computation of a standard error for the Nychka *et al.* (1992) Lyapunov exponent estimate]. It is possible, however, to produce sensitivity plots that are informative about precision, as the ones in Figure 17.1. Figure 17.1 indicates the sensitivity of the dominant Lyapunov exponent estimate to variations in the parameters, by plotting the estimated dominant Lyapunov exponent for each setting of (L, m, k), where $L = 1, 2, 3$, $m = 1, \ldots, 10$, and $k = 1, 2, 3$.

TABLE 17.6
EGARCH (1,1) PARAMETER ESTIMATES AND ERROR TERM DIAGNOSTICS

$$z_t = b_0 + \sum_{j=1}^{q} b_j z_{t-j} + \varepsilon_t, \quad \varepsilon_t \mid I_{t-1} \sim N(0, \sigma_t^2), \quad \log \sigma_t^2 = w_0 + \beta \log(\sigma_{t-1}^2) + \alpha \left| \frac{\varepsilon_{t-1}}{\sigma_{t-1}} \right| + \gamma \frac{\varepsilon_{t-1}}{\sigma_{t-1}}$$

Series	AR Lag, q	EGARCH (1,1) Parameter Estimates				EGARCH (1,1) Error Term Diagnostics (p-values)					
		w_0	α	γ	β	Q-statistic	$Q(\varepsilon^2)$	ARCH	J-B	Log L	$\beta = 1$
C2	1	-1.068 (1.8)	0.347 (3.0)	-0.014 (0.1)	0.881 (10.9)	0.859	0.971	0.918	0.000	284.824	0.153
C3	2	-13.513 (22.9)	-0.004 (0.0)	-0.239 (2.1)	-0.900 (18.5)	0.106	0.983	0.975	0.000	299.714	0.000
nC4	1	-1.083 (1.6)	0.290 (1.5)	0.085 (1.1)	0.809 (6.3)	0.096	0.947	0.937	0.000	130.201	0.139
iC4	2	-6.566 (2.3)	0.475 (2.7)	-0.119 (0.9)	0.185 (0.5)	0.036	0.339	0.799	0.000	337.415	0.030
C5	2	-1.225 (1.4)	0.265 (1.4)	-0.102 (1.1)	0.797 (5.2)	0.609	0.998	0.818	0.000	160.842	0.192
Crude oil	2	-1.220 (2.6)	0.589 (2.0)	-0.063 (0.5)	0.858 (12.9)	0.105	0.999	0.764	0.001	178.376	0.035
Natural gas	3	-1.543 (5.8)	1.271 (4.6)	-0.391 (2.5)	0.812 (26.1)	0.000	0.397	0.288	0.000	10.855	0.000

Notes: Numbers in parentheses next to the EGARCH (1,1) parameter estimates are absolute t-ratios. The Q-statistic is distributed as a $\chi^2(36)$ on the null of no autocorrelation. The ARCH statistic is distributed as a $\chi^2(1)$ on the null of no ARCH. The Jarque-Bera test statistic is distributed as a $\chi^2(2)$ under the null hypothesis of normality.

TABLE 17.7
COMPARISON OF PREDICTIVE POWER FOR THE CONDITIONAL VARIANCE OF BELVIEW ENERGY PRICES

$$\hat{\varepsilon}_t^2 = b_0 + b_1\hat{\sigma}_{t-1}^2 + \zeta_t$$

Series	GARCH (1,1) Results				EGARCH (1,1) Results			
	b_0	b_1	R^2	Q-statistic	b_0	b_1	R^2	Q-statistic
C2	0.000 (0.6)	0.803 (0.8)	0.064	0.512	0.000 (0.5)	0.823 (0.7)	0.064	0.558
C3	0.000 (0.1)	1.119 (0.2)	0.016	0.923	-0.000 (0.5)	1.302 (0.6)	0.046	0.877
nC4	0.002 (0.5)	0.796 (0.7)	0.045	0.977	0.002 (0.5)	0.818 (0.6)	0.045	0.960
iC4	0.000 (0.6)	0.692 (0.7)	0.019	0.631	0.000 (1.1)	0.590 (1.3)	0.026	0.670
C5	-0.001 (0.2)	1.126 (0.2)	0.031	0.998	-0.001 (0.3)	1.173 (0.4)	0.045	0.999
Crude oil	0.004 (2.2)	0.349 (4.5)	0.041	0.999	0.003 (1.5)	0.528 (2.5)	0.053	0.999
Natural gas	0.118 (0.8)	0.302 (13.3)	0.195	0.392	0.150 (1.5)	0.363 (48.3)	0.846	0.969

Notes: Absolute t-statistics for $b_0 = 0$ and $b_1 = 1$ are in parentheses. R^2 is the coefficient of determination. $Q(36)$ is the Ljung-Box statistic for 36 lags of the residual autocorrelation.

TABLE 17.8

The Nychka et al. (1992) BIC Selection of the
Parameter Triple (L, m, k), the Value of the Minimized BIC,
and the Dominant Lyapunov Exponent Point Estimate

Series	(L, m, k) Triple that Minimizes the BIC	Value of the Minimized BIC	Dominant Lyapunov Exponent Point Estimate
C2	(3,3,2)	1.447	0.056
C3	(2,6,2)	1.292	0.211
nC4	(1,7,2)	1.366	0.081
iC4	(2,6,2)	1.386	0.100
C5	(1,4,2)	1.362	0.068
Crude oil	(1,2,1)	1.427	-1.835
Natural gas	(2,8,1)	1.391	-0.063

Note: Numbers in parentheses represent the BIC selection of the parameter triple, (L, m, k), where L is the time delay parameter, m is the number of lags in the autoregression and k is the number of units in the hidden layer of the neural net.

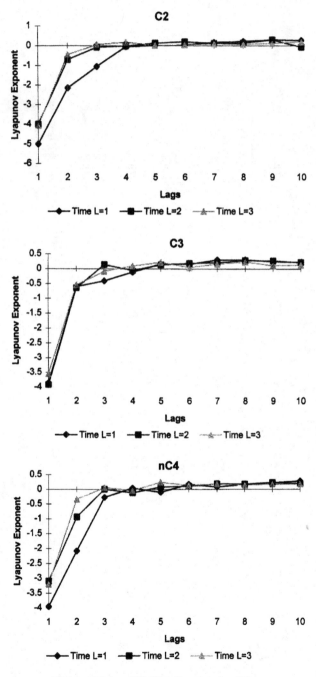

Figure 17.1: NEGM Sensitivity Plots

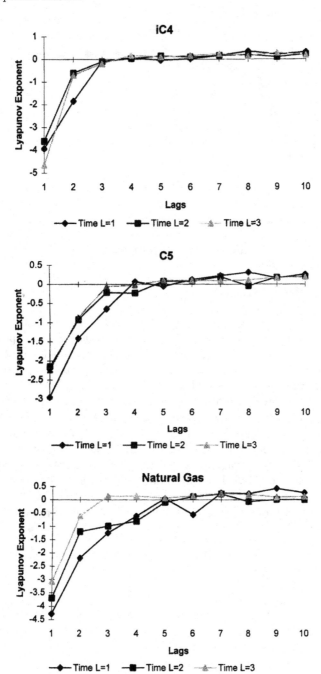

Figure 17.1: (Continued)

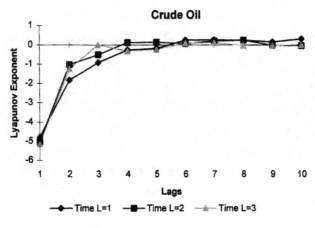

Figure 17.1: (Continued)

17.6   Conclusion

We have provided results of nonlinear dynamical analysis of North American hydrocarbon prices using the Nychka *et al.* (1992) test for positivity of the dominant Lyapunov exponent. Before conducting such a nonlinear analysis, the data were rendered stationary and appropriately filtered, in order to remove any linear as well as nonlinear stochastic dependence.

We have found evidence of nonlinear chaotic dynamics in all five (C2, C3, nC4, iC4, and C5) Belview natural gas liquids markets. In principle, it should be possible to model (by means of differential/difference equations) the nonlinear chaos-generating mechanism and build a predictive model of North American natural gas liquids prices. This is an area for potentially productive future research that will undoubtedly improve our understanding of how North American NGLs prices change over time. See Barnett and Serletis (1999) for more insights regarding this line of research.

Chapter 18

Random Fractal Structures in North American Energy Markets

*Apostolos Serletis and Ioannis Andreadis**

18.1 Introduction

In recent years, the North American energy industry has undergone major structural changes that have significantly affected the environment in which producers, transmission companies, utilities and industrial customers operate and make decisions. For example, major policy changes are the U.S. Natural Gas Policy Act of 1978, Natural Gas Decontrol Act of 1989, and FERC Orders 486 and 636. In Canada, deregulation in the mid-1980s has also broken the explicit link between the delivered prices of natural gas and crude oil (that was in place prior to 1985), and has fundamentally changed the environment in which the Canadian oil and gas industry operates. Moreover, the Free Trade Agreement (FTA) signed in 1988 by the United States and Canada, and its successor, the North American Free Trade Agreement (NAFTA) signed in 1993 by the United States, Canada, and Mexico, have underpinned the process of deregulation and attempted to increase the efficiency of the North American energy industry.

The main objective of this chapter is to follow Serletis and Andreadis

---

*Originally published in *Energy Economics* 26 (2004), 389-399. Reprinted with permission.

(2004) and use tools from dynamical systems theory to explain the price fluctuations in North American crude oil and natural gas markets, using daily data over the deregulated period from the early 1990s to 2001. In this regard, a voluminous literature has developed supporting the efficient markets hypothesis — see, for example, Fama (1970). Briefly stated, the hypothesis claims that asset prices are rationally related to economic realities and always incorporate all the available information, implying the absence of exploitable excess profit opportunities. However, despite the widespread allegiance to the notion of market efficiency, a number of recent studies have suggested that certain asset prices are not rationally related to economic realities — see, for example, Summers (1986) and Serletis and Gogas (1999).

Our principal concern is to distinguish between deterministic and stochastic origin for West Texas Intermediate (WTI) crude oil prices and Henry Hub natural gas prices. In doing so, we implicitly assume that the WTI crude oil price at Chicago is a North American crude oil price (or even a world price) and that the Henry Hub natural gas price at Louisiana is a North American natural gas price — see Serletis and Rangel-Ruiz (2004) for more details. We provide evidence that both WTI crude oil prices and Henry Hub natural gas prices can be explained in the framework of a random fractal time series.

The chapter is organized as follows. Section 18.2 describes the data and investigates their statistical properties. In Sections 18.3 and 18.4 we test for a random multifractal structure, and in Section 18.5 for turbulent behavior. The final section provides a brief conclusion.

18.2   Data and Statistical Analysis

The data consist of daily observations on West Texas Intermediate (WTI) crude oil prices over the period from January 2, 1990 to February 28, 2001 (a total of 2809 observations) and Henry Hub natural gas prices over the period from January 24, 1991 to February 28, 2001 (a total of 2521 observations). Figures 18.1 and 18.2 provide a graphical representation of these series.

18.2.1   The Above and Below Test for Randomness

To investigate the time series properties of these variable, we use the above and below test for randomness — see Spiegel (1988). Let's denote by $T(i)$, $i = 1, ...N$, a time series. The median of the elements of the time series, $T(i), i = 1, ...N$ is denoted by M. We construct a sequence of three symbols, denoted by -1, 0, and $+1$, depending on whether an element of the time

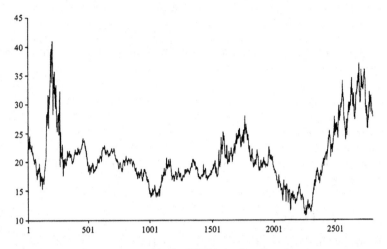

Figure 18.1: Daily WTI crude oil prices.

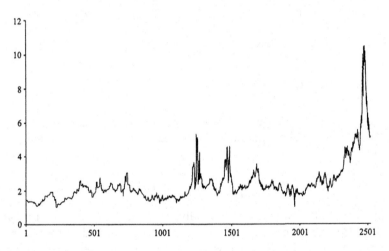

Figure 18.2: Daily Henry Hub natural gas prices.

series is less than, equal to, or greater than M. We denote by N_1 the cardinality of the set of elements $\{+1\}$ and by N_2 that of the set of elements $\{-1\}$. With V we denote the total sign changes from $\{-1\}$ to $\{+1\}$. Then we test whether the sequences of elements N_1, N_2 satisfy a Gaussian distribution.

In doing so, we calculate the mean value and the variance of a binomial distribution for N_1 and N_2 using

$$m_V = \frac{2N_1 N_2}{N_1 + N_2}, \qquad s_V^2 = \frac{2N_1 N_2 (2N_1 N_2 - N_1 - N_2)}{(N_1 + N_2)(N_1 + N_2 - 1)},$$

and construct a variable z following a Gaussian distribution with mean value m_V and variance s_V^2

$$z = \frac{V - m_V}{s_V}.$$

In the case where $-1.96 \leq z \leq 1.96$ there is a randomness behavior for the time series T with a confidence interval of 95%. Applying this test, we find $z = -0.93$ for the WTI oil series and $z = -0.94$ for the Henry Hub natural gas price series.

18.2.2   The Hurst Test

Here we apply the Rescale Range analysis, or Hurst test — see Mandelbrot (1982) and Papaioannou and Karytinos (1995) — which can be described as follows. First, we briefly recall the test. Consider the time series $T(i), i = 1, ..., N$ and for every n, $2 \leq n \leq N$, denote by M_n the mean value of the truncated first n elements. Then we define a new time series $X(j)$ representing the cumulative deviation over the n periods, with elements

$$X(j) = \sum_{j=1}^{n} [T(j) - M_n], \quad j = 1, 2, \cdots, N$$

The range of the cumulative deviation from the average level, R_n, is the difference between the maximum and minimum cumulative deviations over n periods

$$R_n = \max_{1 \leq j \leq n} X(j) - \min_{1 \leq j \leq n} X(j)$$

The function R_n increases with n. Finally, we denote with S_n the standard deviation of the first n elements of the time series T. According to the Hurst law, in the case of a fractional Brownian motion, the following should hold, in the limit of large n

$$\frac{R_n}{S_n} \propto \left(\frac{n}{2}\right)^H,$$

with $0 \leq H \leq 1$ being the Hurst exponent. Hence, we can plot R_n/S_n against $\log(n/2)$ and find a value of the Hurst exponent.

Applying this test in Figures 18.3 and 18.4, we find $H = 0.85$ for crude oil and $H = 0.86$ for natural gas, supporting for both time series a persistent ($H > 0.5$) fractal structure with a long memory.

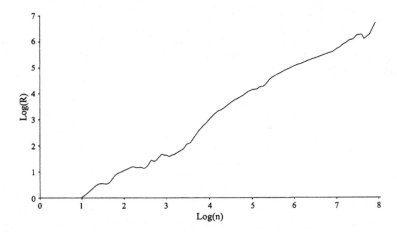

Figure 18.3: The Hurst test for the WTI crude oil price series.

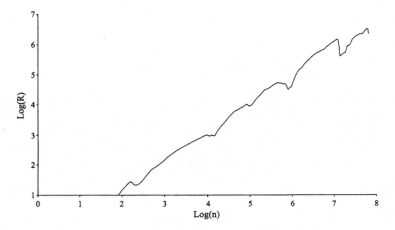

Figure 18.4: The Hurst Test for the Henry Hub natural gas price series.

18.3   A Fractal Noise Model

In recent years, a significant volume of research supports the existence of nonlinear dynamics in most economic and financial time series. Nonlinearity, however, could be either deterministic or stochastic — see, for example, Barnett and Serletis (2000). In this section, we provide evidence that the nonlinearity in the WTI crude oil and Henry Hub natural gas price series has a noise origin.

18.3.1   The Power Spectrum

As we have time series with a finite number of data points, we follow Li (1991) and calculate its power spectrum $P(f)$, using the following discrete Fourier transform

$$P(f) = N\|A(f)\|^2, \tag{18.1}$$

where $\|A(f)\|$ is the module of the complex number

$$A(f) = \frac{1}{N} \sum_{j=1}^{N} x_j e^{\frac{i2\pi f j}{N}}. \tag{18.2}$$

We present the power spectrum of the crude oil and natural gas time series in Figures 18.5 and 18.6, respectively. We find behavior of the type $1/f^\alpha$, where $\alpha = 2.03$ for crude oil and $\alpha = 1.78$ for natural gas. This behavior is strictly related to the self-critical phenomena reported by Bak and Chen (1991), and is consistent with the evidence reported in Andreadis (2000) for the S&P 500.

18.3.2   The Structure Function Test

Next we apply the structure function test, developed by Provenzale *et al.* (1992), in order to support and extend the results obtained so far indicating a fractal noise model. The structure function test was originally developed as a tool for distinguishing between a deterministic and a stochastic origin of time series whose power spectrum displays a scaling behavior.

We consider a time series T with a finite length equal to N. For every n, $1 \leq n \leq N$, the structure function associated with T is defined as follows

$$\Sigma(n) = \sum_{i=1}^{N-n} \Big[T((i+n)\,\Delta t) - T(i\Delta t) \Big]^2, \tag{18.3}$$

where Δt denotes the sampling rate of T. According to Mandelbrot (1982), for a time series T with a power-law spectrum $P(f) \propto f^{-\alpha}$, where α is positive real, one expects a scaling behavior of the form $\Sigma(n) \propto n^{2H}$ at small values of n, where H is called the scaling exponent. In the case of a fractional Brownian motion, it holds [see Provenzale *et al.* (1992)] that

$$\alpha = 2H + 1. \tag{18.4}$$

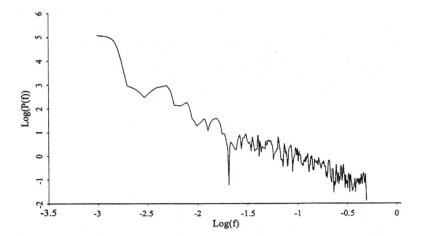

Figure 18.5: The power spectrum of the WTI crude oil series.

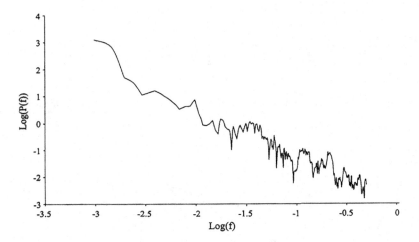

Figure 18.6: The power spectrum of the Henry Hub natural gas series.

When the signal is a fractal noise, the graph of $\log(\Sigma(n))$ versus $\log(n)$ displays an extended scaling regime and it is closely approximated by a straight line. On the other hand, if the time series corresponds to the motion of a strange attractor whose fractal structure is due to close returns in phase space, the graph of $\log(\Sigma(n))$ versus $\log(n)$ is closely approximated at small values of n, by a straight line with slope $2H = 1$. At intermediate n, $\Sigma(n)$ has an oscillatory behavior, due to orbit occurrence in phase space.

Finally, for high values of n, $\Sigma(n)$ approaches a constant, due to the limited phase space visited by the system.

In Figures 18.7 and 18.8, we show the graph of $\log(\Sigma(n))$ versus $\log(n)$ for the WTI crude oil and Henry Hub natural gas price series. We find that $H = 0.69$ for crude oil and $H = 0.56$ for natural gas. In both cases, $\alpha > 2$, rejecting a fractional Brownian motion and supporting behavior like a fractal noise.

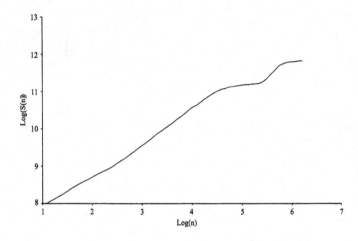

Figure 18.7: The structure function test for the WTI crude oil series.

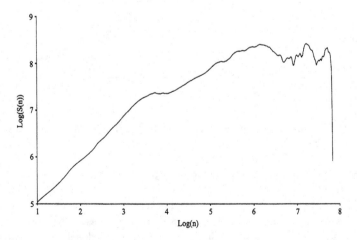

Figure 18.8: The structure function test for the Henry Hub natural gas series.

18.4　A Multifractal Data Analysis

Let us consider the time series $T(i)$, $i = 1, 2, ..., N$. Vassilicos *et al.* (1993) addressed the question of whether the set of points $T(i)$, $i = 1, 2, ..., N$, is a fractal set on the time axis, in the sense of the presence of a scaling. If the answer is yes, then the next interesting question is whether this fractal distribution is homogeneous or whether it is a multifractal, in the sense that the different fractal scalings may apply to different times.

To answer this question in the context of the WTI crude oil and Henry Hub natural gas price series, we calculate the generalized dimensions D_q of the graph of the time series $T(i)$. Let us briefly recall this algorithm. The time axis is covered by a grid of points separated with a fixed distance ε from each other. We label each interval between grid points with an integer j and calculate the total number of announcements, μ_j, that lie within the interval j. Then we compute the quantity

$$N_q(\varepsilon) = \sum_j \mu_j^q.$$

and for various integers we use $q = 0, 2, 3, 4$.

When the distribution of the points is fractal in the sense of Vassilicos and Hunt (1991), then

$$N_q(\varepsilon) = \varepsilon^{-D_0}$$

for ε small enough and $0 < D_0 < 1$. D_0 is called the fractal dimension and characterizes the fractal structure of the set. If there exists a D_0 then there also exist powers τ_q such that

$$N_q(\varepsilon) = \varepsilon^{-\tau_q}$$

for ε small enough and for integer values of q. Note that $\tau_1 = 0$ and the generalized dimension D_q are defined by

$$D_q = \frac{\tau_q}{1 - q}$$

and $1 \geq D_0 \geq D_1 \geq D_2 \geq ... \geq 0$.

We present the numerical results obtained by applying the previous algorithm in Table 18.1.

TABLE 18.1
TESTS FOR MULTIFRACTAL STRUCTURE

	WTI crude oil		Henry Hub natural gas	
q	τ_q	D_q	τ_q	D_q
$q = 0$	0.85	0.85	0.85	0.85
$q = 2$	−0.93	0.93	−0.82	0.82
$q = 3$	−1.89	0.94	−1.72	0.86
$q = 4$	−2.83	0.94	−2.55	0.85

Clearly, these results do not support a multifractal structure for the Henry Hub natural gas price series but they do for the WTI crude oil price series. In particular, in the case of WTI oil the values of the multifractal dimension approach 1.

18.5   On Turbulent Behavior

Recently, Ghashghaie *et al.* (1996) have advanced the hypothesis of turbulent behavior in financial markets. Their hypothesis, however, has been criticized by Mantegna and Stanley (1996). To provide some evidence on this issue, we report a scaling behavior for the WTI crude oil and Henry Hub natural gas price series which agrees with the Ghashghaie *et al.* (1996) hypothesis of turbulent behavior.

Let us define the return over n time steps as $Z_n(t) = |X(t + \Delta t) - X(t)|$, where $X(t)$ is an entry of the time series and $\Delta t = 1$ is the sampling time. We have found that the moments of the distribution $Z_n(t)$ possessing a scaling behavior as a function of n, can be expressed as

$$\langle | Z_n(t) |^q \rangle_t \propto n^{\xi(q)}$$

where $\xi(q)$ is the self-affinity exponent. In Table 18.2, we indicate the values obtained for $q = 1, 2, 3$ for the WTI crude oil and Henry Hub natural gas price series and compare them with the values $\xi(q) = q/3$ for turbulent flows.

TABLE 18.2
TESTS FOR TURBULENCE

q	Turbulent flows	WTI crude oil	Henry Hub gas
$q = 1$	0.33	0.42	0.35
$q = 2$	0.66	0.79	0.62
$q = 3$	1.00	1.01	0.83

Clearly, the behavior of the WTI crude oil price is consistent with the Ghashghaie *et al.* (1996) hypothesis of turbulent behavior. The Henry Hub natural gas price series, however, is not.

18.6   Conclusion

We have used daily observations on WTI crude oil and Henry Hub natural gas prices and applied tests from dynamical systems theory to distinguish between deterministic and stochastic origin for the series. We provide evidence for a random multifractal turbulent structure for WTI crude oil prices, consistent with the evidence reported (for other markets) by Vassilicos *et al.* (1993), Ghashghaie *et al.* (1996) and Ivanova and Ausloos (1999). Henry Hub natural gas prices, however, are only consistent with a random fractal model.

Our results are also consistent with those reported by Serletis and Gogas (1999). Using the Lyapunov exponent estimator of Nychka, Ellner, Gallant, and McCaffrey (1992), they find evidence of nonlinear chaotic dynamics in North American natural gas liquids markets (those of ethane, propane, normal butane, iso-butane, and naptha) but not in the crude oil and natural gas markets.

Chapter 19

Randomly Modulated Periodic Signals in Alberta's Electricity Market

*Melvin Hinich and Apostolos Serletis**

19.1 Introduction

As Bunn (2004, p. 2) recently put it "the crucial feature of price formation in electricity spot markets is the instantaneous nature of the product. The physical laws that determine the delivery of power across a transmission grid require a synchronised energy balance between the injection of power at generating points and the offtake at demand points (plus some allowance for transmission losses). Across the grid, production and consumption are perfectly synchronised, without any capability for storage. If the two get out of balance, even for a moment, the frequency and voltage of the power fluctuates. Furthermore, end-users treat this product as a service at their convenience. When we go to switch on a light, we do not re-contract with a supplier for the extra energy before doing so. We just do it, and there is a tendency for millions of other people to do likewise whenever they feel like. Electricity may be produced as a commodity, but it is consumed

*Originally published in *Studies in Nonlinear Dynamics and Econometrics* 10(3) (2006), Article 5. Reprinted with permission.

as a service. The task of the grid operator, therefore, is to be continously monitoring the demand process and to call on those generators who have the technical capability and the capacity to respond quickly to the fluctuations in demand."

Recent leading-edge research has applied various innovative methods for modeling spot wholesale electricity prices — see, for example, Deng and Jiang (2004), León and Rubia (2004), and Serletis and Andreadis (2004). The main objective of this chapter is to use a parametric statistical model, called Randomly Modulated Periodicity (RMP), recently proposed by Hinich (2000) and Hinich and Wild (2001), to study Alberta's spot wholesale power market, defined on hourly intervals (like most spot markets for electricity are). In doing so, we use hourly electricity prices, denominated in megawatt-hours (MWh), and MWh demand over the recent deregulated period from 1/1/1996 to 12/7/2003 (a total of over 65000 observations, since there are 8760 hours in a normal year). Our principal concern is to test for periodic signals in electricity prices and electricity load — that is, signals that can be perfectly predicted far into the future since they perfectly repeat every period. In doing so, we take a univariate approach, although from an economic perspective the interest in the price of electricity is in its relationship with the electricity load and perhaps with the prices of other primary fuel commodities.

The chapter is organized as follows. In Sections 19.2 and 19.3 we briefly discuss the RMP model, proposed by Hinich (2000) and Hinich and Wild (2001), for the study of varying periodic signals. In Section 19.4 we briefly discuss Alberta's power market and in Section 19.5 we test for randomly modulated periodicity in hourly electricity prices and MWh demand over the deregulated period after 1996. The final section provides a brief conclusion.

19.2 Randomly Modulated Periodicity

All signals that appear to be periodic have some sort of variability from period to period regardless of how stable they appear to be in a data plot. A true sinusoidal time series is a deterministic function of time that never changes and thus has zero bandwidth around the sinusoid's frequency. Bandwidth, a term from Fourier analysis, is the number of frequency components that are needed to have an accurate Fourier sum expansion of a function of time. A single sinusoid has no such expansion. A zero bandwidth is impossible in nature since all signals have some intrinsic variability over time.

Deterministic sinusoids are used to model cycles as a mathematical convenience. It is time to break away from this simplification in order to model the various periodic signals that are observed in fields ranging from biology, communications, acoustics, astronomy, and the various sciences.

Hinich (2000) introduced a parametric statistical model, called Randomly Modulated Periodicity (RMP), that allows one to capture the intrinsic variability of a cycle. A discrete-time random process $x(t_n)$ is an RMP with period $T = N\tau$ if it is of the form

$$x(t_n) = s_0 + \frac{2}{N} \sum_{k=1}^{N/2} [(s_{1k} + u_{1k}(t_n)) \cos(2\pi f_k t_n) + (s_{2k} + u_{2k}(t)) \sin(2\pi f_k t_n)]$$

where $t_n = n\tau$, τ is the sampling interval, $f_k = k/T$ is the k-th Fourier frequency, and where for each period the $\{u_{11}(t_1), \ldots, u_{1,N/2}(t_n),$ $u_{21}(t_n), \ldots, u_{2,N/2}(t_n)\}$ are random variables with zero means and a joint distribution that has the following finite dependence property: $\{u_{jr}(s_1), \ldots, u_{jr}(s_m)\}$ and $\{u_{ks}(t_1), \ldots, u_{ks}(t_n)\}$ are independent if $s_m + D < t_1$ for some $D > 0$ and all $j, k = 1, 2$ and $r, s = 1, \ldots, N/2$ and all times $s_1 < \cdots < s_m$ and $t_1 < \cdots < t_n$. Finite dependence is a strong mixing condition — see Billingsley (1979).

These time series, $u_{k1}(t)$ and $u_{k2}(t)$, are called 'modulations' in the signal processing literature. If $D << N$ then the modulations are approximately stationary within each period. The process $x(t_n)$ can be written as

$$x(t_n) = s(t_n) + u(t_n),$$

where

$$s(t_n) = E[x(t_n)] = s_0 + \frac{2}{N} \sum_{k=1}^{N/2} [s_{1k} \cos(2\pi f_k t_n) + s_{2k} \sin(2\pi f_k t_n)]$$

and

$$u(t_n) = \frac{2}{N} \sum_{k=1}^{N/2} [u_{1k} \cos(2\pi f_k t_n) + u_{2k} \sin(2\pi f_k t_n)]$$

Thus $s(t_n)$, the expected value of the signal $x(t_n)$, is a periodic function. The fixed coefficients s_{1k} and s_{2k} determine the shape of $s(t_n)$. If $s_{11} \neq 0$ or $s_{21} \neq 0$ then $s(t_n)$ is periodic with period $T = N\tau$. If $s_{11} = 0$ and $s_{21} = 0$, but $s_{12} \neq 0$ or $s_{22} \neq 0$, then $s(t_n)$ is periodic with period $T/2$. If the first $k_0 - 1$ s_{1k} and s_{2k} are zero, but not the next, then $s(t_n)$ is periodic with period T/k_0.

19.3   Signal Coherence Spectrum

To provide a measure of the modulation relative to the underlying periodicity, Hinich (2000) introduced a concept called the signal coherence spectrum (SIGCOH). For each Fourier frequency $f_k = k/T$ the value of SIGCOH is

$$\gamma_x(k) = \sqrt{\frac{|s_k|^2}{|s_k|^2 + \sigma_u^2(k)}}$$

where $s_k = s_{1k} + is_{2k}$ is the amplitude of the kth sinusoid written in complex variable form, $i = \sqrt{-1}$, $\sigma_u^2(k) = E|U(k)|^2$ and

$$U(k) = \sum_{n=0}^{N-1} u_k(t_n) \exp(-i2\pi f_k t_n)$$

is the discrete Fourier transform (DFT) of the modulation process $u_k(t_n) = u_{1k}(t_n) + iu_{2k}(t_n)$ written in complex variable form.

Each $\gamma_x(k)$ is in the $(0, 1)$ interval. If $s_k = 0$ then $\gamma_x(k) = 0$. If $U(k) = 0$ then $\gamma_x(k) = 1$. The SIGCOH measures the amount of 'wobble' in each frequency component of the signal $x(t_n)$ about its amplitude when $s_k > 0$. The amplitude-to-modulation standard deviation (AMS) is

$$\rho_x(k) = \frac{|s_k|}{\sigma_u(k)}$$

for frequency f_k. Thus,

$$\gamma_x^2(k) = \frac{\rho_x^2(k)}{\rho_x^2(k) + 1}$$

is a monotonically increasing function of this signal-to-noise ratio. Inverting this relationship, it follows that

$$\rho_x^2(k) = \frac{\gamma_x^2(k)}{1 - \gamma_x^2(k)}$$

An AMS of 1.0 equals a signal coherence of 0.71 and an AMS of 0.5 equals a signal coherence of 0.45.

To estimate the SIGCOH, $\gamma_x(k)$, suppose that we know the fundamental period and we observe the signal over M such periods. The mth period is $\{x((m-1)T + t_n), n = 0, \ldots, N-1\}$. The estimator of $\gamma(k)$ introduced by Hinich (2000) is

$$\hat{\gamma}(k) = \sqrt{\frac{|\bar{X}(k)|^2}{|\bar{X}(k)|^2 + \hat{\sigma}_u^2(k)}},$$

where

$$\bar{X}(k) = \frac{1}{M} \sum_{m=1}^{M} X_m(k)$$

is the sample mean of the DFT,

$$X_m(k) = \sum_{n=0}^{N-1} x((m-1)T + t_n) \exp(-i2\pi f_m t_n),$$

and

$$\hat{\sigma}_u^2(k) = \frac{1}{M} \sum_{m=1}^{M} |X_m(k) - \bar{X}(k)|^2$$

is the sample variance of the residual discrete Fourier transform, $X_m(k) - \bar{X}(k)$. This estimator is consistent as $M \to \infty$ and if the modulations have a finite dependence of span D then the distribution of

$$Z(k) = \frac{M}{N} \frac{|\bar{X}(k)|^2}{\sigma_u^2(k)}$$

is asymptotically chi-squared with two degrees-of-freedom and a noncentrality parameter $\lambda_k = (M/N)\rho_x^2(k)$ as $M \to \infty$ — see Hinich and Wild (2001). These $\chi_2^2(\lambda_k)$ variates are approximately independently distributed over the frequency band when $D << N$.

If the null hypothesis for frequency f_k is that $\gamma_x(k) = 0$ and thus its AMS is zero, then $Z(k)$ is approximately a central chi-squared statistic. Thus $Z(k)$ can be used to falsify the null hypothesis that $\gamma_x(k) = 0$. The tests across the frequency band are approximately independently distributed tests. The use of the transformation to the $Z(k)$'s is the only straightforward way to put statistical confidence on the signal coherence point estimates.

19.4   Alberta's Power Market

Electricity demand in Alberta is comprised of four primary groups: residential, farm, commercial, and industrial. Alberta has unique load requirements compared with other North American power markets. In particular, the industrial load is over 50% of all electric sales while the residential load is only 15%. This provides a very stable load curve all over the year, which helps reduce the frequency of price spikes. The main contributors to the fluctuations of demand are residential and small commercial customers.

Electricity demand is also cyclical in nature, with demand being lower in the spring and fall than in summer and winter. In fact, Alberta has higher winter consumption, due to lower temperatures that cause increased heating and shorter daylight hours. Moreover, winter hourly load in Alberta has two distinct peaks. Demand is low in the early morning hours and begins to increase through the morning hours, with a first peak around nine o'clock. During this interval, load can increase by 1500 MW. The other peak is around dinnertime, six-seven o'clock. The peaks in the day tend to float depending on the number of daylight hours. Demand also follows a weekly cycle and tends to be higher on weekdays than during the weekends.

Finally, demand for power is relatively inelastic in Alberta. There is no requirement for load to bid in the price they are willing to pay for the energy. Hence, un-bid load is treated as a price taker into the merit order of the Alberta's power market and must pay the hourly pool price for the energy consumed during that hour. Only a small percentage (around 4%) of the load is bid into the market. According to AESO Operations group, there is around 300 MW of price responsive demand: some large industrial customers agree to be curtailed at high pool prices, introducing some price sensitivity at higher price levels.

Being a deregulated market, the pool price in Alberta's power market is determined by competitive market forces; that is, the laws of supply and demand. Being components of supply and demand, imports and exports also influence electricity prices. Import and export volumes play an important role in ensuring system reliability and security in Alberta. In conditions of scarcity of supply and/or excess of demand, power must be imported via the inter tie-lines that connect the Alberta electric grid system with the neighbouring jurisdictions. In fact, the Alberta Interconnected Electric System (AIES) is connected, on the west side, to the British Columbia (BC) grid by the 800 MW Alberta-BC inter-tie, while it is linked on the east side to the Saskatchewan power system by a 150 MW DC interconnection.

Since the total available capacity of the inter-ties represents about 11% of the Alberta maximum peak load, the tie lines work as very large generating units, and thus may have a considerable impact on the pool price. This fact, in combination with Alberta's steep supply curve and inelastic demand even at high prices, has given importers and exporters significant market power, which has raised concerns among market participants. In conditions of tight supply-demand balance, the pool price is strongly impacted by the discretionary sales tactics implemented by importers. While these strategies of adjusting the volume of imports and exports in response to market outcomes are normal profit-maximizing behaviours, on the other hand, practices like abuse of market power or electricity dumping are deemed to manipulate the pool price. These undesirable practices have been the issues of a controversy among stakeholders in the Alberta's electricity market — see Bianchi and Serletis (2006) for more details.

19.5   RMP in Alberta's Power Market

We use the time series of Alberta hourly spot prices and megawatt-hours (MWh) demand from 1/1/1996 to 12/7/2003. Figure 19.1 shows a section of the demand time series (the load curve) and Figure 19.2 shows a section of the spot prices time series, over the same period. Electricity demand has a daily and weekly cycle but it is clear from Figure 19.1 that these cycles in the demand are wobbly. However, it is hard to see a daily or weekly cycle in the spot electricity prices in Figure 19.2. The prices time series in Figure 19.2 is much more spiky, shows higher volatility, and also a stronger mean-reverting pattern than the load time series in Figure 19.1.

We applied the signal coherence spectral analysis to the time series of spot prices and demand, using the FORTRAN 95 'Spectrum.for' program developed by Hinich and available at his web page, www.la.utexas.edu/~hinich. In doing so, we first detrended the hourly electricity demand and the hourly spot electricity price data by fitting an AR(12) model to each series — the AR(12) filter is used to make the data have a flat spectrum; it is a linear transformation and thus it does not create nor destroy coherence. The residuals of the fitted model are then analyzed for the presence of a randomly modulated periodicity with a fundamental period of one week (168 hours). An AR fit is a linear operation that cannot create signal coherence. Indeed signal coherence can only be reduced by a improperly applied detrended method.

The adjusted R square for the demand data is 0.74. The characteristic polynomial of the estimated AR(12) model has a 4th order complex root pair whose amplitudes are 0.96 and a complex root pair whose amplitudes are 0.96. The amplitude of the other root pair is 0.71. The adjusted R square for the spot price data is 0.666. The largest root magnitude is 0.81.

The SIGCOH spectrum of the demand time series is shown in Figure 19.3. All the long period harmonics up to the period of 8.4 hours have coherence greater than 0.5 except for the 9.99 hour harmonic. Only the fundamental and the harmonics 28 and 24 hours have coherences greater than 0.9. The shorter period components are either not very coherent or incoherent. Figure 19.3 also shows the conventional power spectrum (log spectrum in decibels). The harmonic peaks in the spectrum indicate that the weekly and daily cycles are not simple sinusoids but their lack of amplitude and phase stability indicated in the SIGCOH spectrum implies that the shorter period components are of little use for forecasting.

The SIGCOH spectrum of the spot prices is shown in Figure 19.4. Only the 24 hour harmonic has a coherence barely greater than 0.75. The rest have coherences less than 0.5, including the fundamental. The plot of the power spectrum in Figure 19.4 shows the standard methods for fitting a

Fourier expansion of the weekly and daily cycles will not contribute much to a forecast of the spot prices.

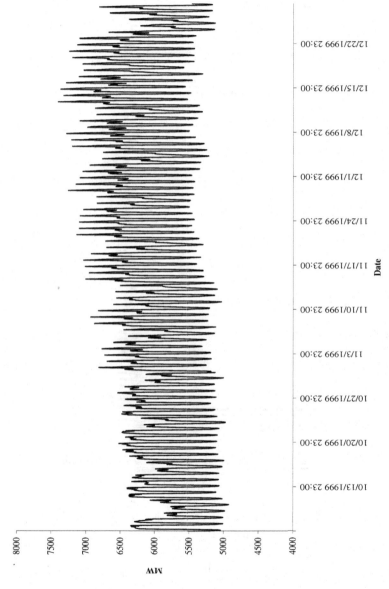

Figure 19.1: A Section of Alberta Electricity Demand

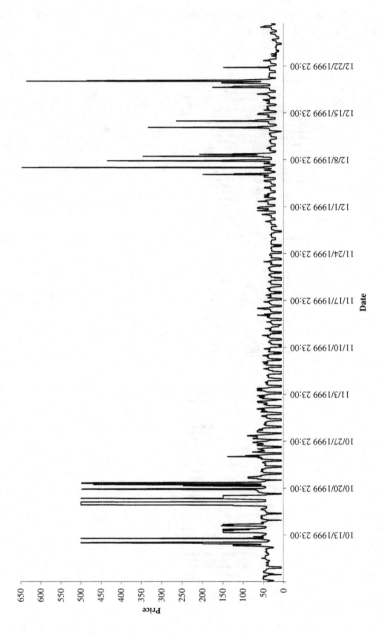

Figure 19.2: A Section of Alberta Electricity Spot Prices

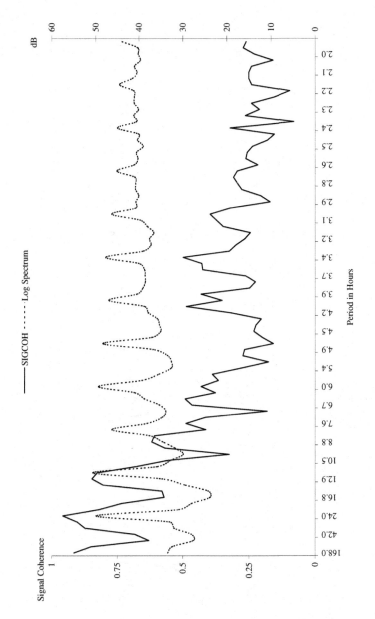

Figure 19.3: Power & Signal Coherence Spectra of the Residuals from an AR(12) Fit of the Alberta Electricity Hourly Spot Demand

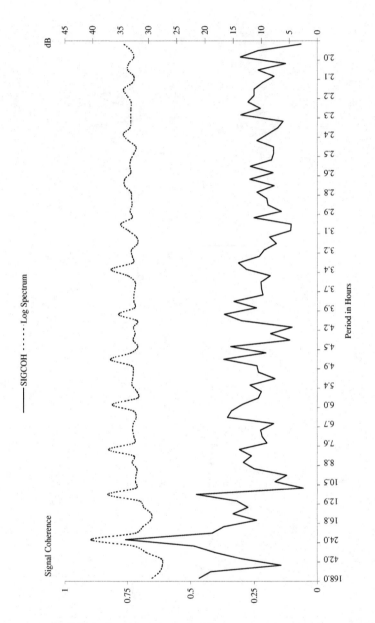

Figure 19.4: Power & Signal Coherence Spectra of the Residuals from an AR(12) Fit of the Alberta Electricity Hourly Spot Prices

19.6 Conclusion

We have applied the signal coherence spectral analysis to the time series of hourly spot prices and megawatt-hours (MWh) demand for Alberta and found that electricity prices have low coherence in the daily and weekly cycles, meaning that forecast errors will have a high error variance. However, electricity demand has lot of high coherence with the daily and weekly cycles being stable with some variation. The mean values at each half hour of the daily demand and the weekend demand should yield good forecasts for a day and the weekend for the next week after the end of the data series. Yet we expect that a statistical forecasting based on the historical demand and cofactors such as the average hourly temperature per day and patterns of industrial usage should yield better short term forecasts. Clearly, the development of a statistical technology for forecasting electricity demand is an exciting and challenging area of research — see, for example, Li and Hinich (2002).

In this chapter, we have taken a univariate time series approach to the analysis of electricity prices. From an economic perspective, however, the interest in the price of electricity is in its relationship with the prices of various underlying primary fuel commodities such as, for example, natural gas, oil, or coal. As Bunn (2004, p. 2) recently put it

> "··· take the case of gas, for example. This is now becoming the fuel of choice for electricity generation. The investment costs are lower than coal, or oil plant; it is cleaner and, depending upon location, the fuel costs are comparable. But with more and more of the gas resources being used for power generation, in some markets the issue of whether gas drives power prices, or *vice versa*, is not easily answered."

Because the properties of univariate series need not be at all like the properties of their multivariate relationships, investigating the relationship between electricity prices and the prices of other primary fuel commodities is an area for potentially productive future research.

Bibliography

[1] Alberta Electric System Operator. "Alberta Import / Export Tariff." (September 23, 2004).

[2] Alberta Department of Energy. "Alberta's Electricity Policy Framework: Competitive – Reliable – Sustainable." (June 6, 2005).

[3] Alberta Market Surveillance Administrator *Report*. "Zero Dollar Offers." (April 29, 2003).

[4] Alberta Market Surveillance Administrator *Report*. "A Review of Imports, Exports, and Economic Use of the BC Interconnection." (January 10, 2005).

[5] Alberta Market Surveillance Administrator *Report*. "Update on Economic Use of the BC Interconnection." (September 23, 2005).

[6] Andreadis, I. "Self-Criticality and Stochasticity of a S&P 500 Index Time Series." *Chaos, Solitons, and Fractals* 11 (2000), 1047-1059.

[7] Anderson, R.W. "Some Determinants of the Volatility of Futures Prices." *Journal of Futures Markets*, 5 (1985), 331-348.

[8] Bailey, E. *The Geographic Expanse of the Market for Wholesale Electricity*. MIT unpublished Ph.D. thesis (1998).

[9] Baillie, R.T. and T. Bollerslev. "Common Stochastic Trends in a System of Exchange Rates." *Journal of Finance* 44 (1989), 167-181.

[10] Bak, P. and K. Chen. "Self-Organized Criticality." *Scientific American* 264 (1991), 26-33.

[11] Barnett, W.A. and A. Serletis. "Martingales, Nonlinearity, and Chaos." *Journal of Economic Dynamics and Control* 24 (2000), 703-724.

[12] Barnett, W.A., A.R. Gallant, M.J. Hinich, J.A. Jungeilges, D.T. Kaplan, and M.J. Jensen. "Robustness of Nonlinearity and Chaos Test to Measurement Error, Inference Method, and Sample Size." *Journal of Economic Behavior and Organization* 27 (1995), 301-320.

[13] Barnett, W.A., A.R. Gallant, M.J. Hinich, J.A. Jungeilges, D.T. Kaplan, and M.J. Jensen. "A Single-Blind Controlled Competition Among Tests for Nonlinearity and Chaos." *Journal of Econometrics* 82 (1997), 157-192.

[14] Baxter, M. and R.G. King. "Approximate Band-pass Filters for Economic Time Series." NBER Working Paper No. 5052 (1995).

[15] Baxter, M. and R.G. King "Measuring Business Cycles: Approximate Band-Pass Filters for Economic Time Series." *The Review of Economics and Statistics* 81 (1999), 575-593.

[16] Bera, A.K. and C.M. Jarque. "Efficient Tests for Normality, Heteroscedasticity, and Serial Independence of Regressions." *Economics Letters* 6 (1980), 255-259.

[17] Bernard, A.B. "Empirical Implications of the Convergence Hypothesis." Working Paper, MIT, Cambridge, MA. (1992).

[18] Beveridge, S. and C.R. Nelson. "A New Approach to Decomposition of Economic Time Series into Permanent and Transitory Components with Particular Attention to Measurement of the Business Cycle." *Journal of Monetary Economics* 7 (1981), 151-174.

[19] Billingsley, P. *Probability and Measure.* New York: John Wiley (1979).

[20] Bollerslev, T. "Generalized Autoregressive Conditional Heteroscedasticity." *Journal of Econometrics* 31 (1986), 307-327.

[21] Bollerslev, T., R. Chou, and K. Kroner. "ARCH Modeling in Finance: A Review of the Theory and Empirical Evidence." *Journal of Econometrics* 52 (1992), 5-59.

[22] Brennan, M.J. "The Supply of Storage." *American Economic Review* 48 (1958), 50-72.

[23] Brock, W.A., W.D. Dechert, J. Scheinkman, and B. LeBaron "A Test for Independence Based on the Correlation Dimension." *Econometric Reviews* 15 (1996), 197-235.

[24] Bunn, D.W. "Structural and Behavioural Foundations of Competitive Electricity Prices." In Derek W. Bunn (ed.), *Modelling Prices in Competitive Electricity Markets*, Wiley Series in Financial Economics (2004), 1-17.

[25] Burns, A.F. and W.C. Mitchell. *Measuring Business Cycles*. New York: NBER (1946).

[26] Canadian Energy Research Institute. "Electric Power Industry Fundamentals & the Role of The AESO." (2004).

[27] Cho, D.W. and G.S. McDougall. "The Supply of Storage in Energy Futures Markets." *The Journal of Futures Markets* 10 (1990), 611-21.

[28] Christian, J. and K. Hughes. "The BC-Alberta Intertie: Impact of Regulatory Change." (June 16, 2004).

[29] Christiano, L.J. "Searching for a Break in GNP." National Bureau of Economic Research Working Paper 2695 (1988).

[30] Christiano, L.J. and M. Eichenbaum. "Unit Roots in GNP: Do We Know and Do We Care?" NBER Working Paper, No. 3130 (1989).

[31] Cochrane, J.H. "How Big is the Random Walk in GNP?" *Journal of Political Economy* 96 (1988), 893-920.

[32] Cochrane, J.H. "A Critique of Unit Root Tests." *Journal of Economic Dynamics and Control* 15 (1991), 275-284.

[33] Coe, P. and A. Serletis. "Bounds Tests of the Theory of Purchasing Power Parity." *Journal of Banking and Finance* 26 (2001), 179-199.

[34] Cogley, T. "International Evidence on the Size of the Random Walk in Output." *Journal of Political Economy* 98 (1990), 501-518.

[35] Cogley, T. and J.M. Nason. "Effects of the Hodrick-Prescott Filter on Trend and Difference Stationary Time Series: Implications for Business Cycle Research". *Journal of Economic Dynamics and Control* 19 (1995), 253-278.

[36] Czamanski, D., P. Dormaar, M.J. Hinich, and A. Serletis. "Episodic Nonlinearity and Nonstationarity in Alberta's Power and Natural Gas Markets." *Energy Economics* (forthcoming, 2006).

[37] Day, T. and C. Lewis. "Forecasting Futures Market Volatility." *Journal of Derivatives* 1 (1993), 33-50.

[38] Day, T. and C. Lewis. "Initial Margin Policy and Stochastic Volatility in the Crude Oil Futures Market." *Review of Financial Studies* 10 (1997), 303-332.

[39] Deaves, R. and I. Krinsky. "Risk Premiums and Efficiency in the Market for Crude Oil Futures." *The Energy Journal* 13 (1992), 93-117.

[40] Deng, S.J. and W. Jiang. "Quantile-Based Probabilistic Models for Electricity Prices." In Derek W. Bunn (ed.), *Modelling Prices in Competitive Electricity Markets*, Wiley Series in Financial Economics) (2004), 161-176.

[41] De Vany, A. and W.D. Walls. "Pipeline Access and Market Integration in the Natural Gas Industry: Evidence from Cointegration Tests." *The Energy Journal* 14 (1993), 1-19.

[42] De Vany, A. and W.D. Walls. "Cointegration Analysis of Spot Electricity Prices: Insights on Transmission Efficiency in the Western US." *Energy Economics* 21 (1999a), 435-448.

[43] De Vany, A. and W.D. Walls. "Price Dynamics in a Network of Decentralized Power Markets." *Journal of Regulatory Economics* 15 (1999b), 123-140.

[44] Dickey, D.A. and W.A. Fuller. "Distribution of the Estimators for Autoregressive Time Series with a Unit Root." *Journal of the American Statistical Association* 74 (1979), 427-431.

[45] Dickey, D.A. and W.A. Fuller. "Likelihood Ratio Statistics for Autoregressive Time Series with a Unit Root." *Econometrica* 49 (1981), 1057-72.

[46] Diks, C. and V. Panchenko. "A New Statistic and Practical Guidelines for Nonparametric Granger Causality Testing." Mimeo, Department of Economics, University of Amsterdam, (2005b).

[47] Doldado, J., T. Jenkinson, and S. Sosvilla-Rivero. "Cointegration and Unit Roots." *Journal of Economic Surveys* 4 (1990), 249-273.

[48] Eckmann, J.P. and D. Ruelle. "Ergodic Theory of Strange Attractors." *Reviews of Modern Physics* 57 (1985), 617-656.

[49] Ellner, S., D.W. Nychka, and A.R. Gallant. "LENNS, a Program to Estimate the Dominant Lyapunov Exponent of Noisy Nonlinear Systems from Time Series Data." Institute of Statistics Mimeo Series

#2235 (BMA Series #39), Statistics Department, North Caroline State University, Raleigh, NC 27694-8203, (1992).

[50] Engle, R.F. "Autoregressive Conditional Heteroscedasticity with Estimates of the Variance of U.K. Inflation." *Econometrica* 50 (1982), 987-1008.

[51] Engle, R.F. and C.W. Granger. "Cointegration and Error Correction: Representation, Estimation and Testing." *Econometrica* 55 (1987), 251-276.

[52] Engle, R.F. and S. Kozicki. "Testing for Common Features." *Journal of Business and Economic Statistics* 11 (1993), 369-380.

[53] Engle, R.F. and B.S. Yoo. "Forecasting and Testing in Cointegrated Systems." *Journal of Econometrics* 35 (1987), 143-159.

[54] Engle, R.F. and B.S. Yoo. *"Cointegrated Economic Time Series: A Survey with New Results."* UCSD Discussion Paper 87-26R (1989).

[55] Ericsson, N.R. "Comment." *Journal of Business and Economic Statistics* 11 (1993), 380-383.

[56] Fama, E.F., "Efficient Capital Markets: A Review of Theory and Empirical Work." *Journal of Finance* 25 (1970), 383-417.

[57] Fama, E.F. "Forward and Spot Exchange Rates." *Journal of Monetary Economics* 14 (1984), 319-338.

[58] Fama, E.F. and K.R. French. "Business Cycles and the Behavior of Metals Prices." *Journal of Finance* 43 (1988), 1075-1093.

[59] Fiorito, R. and T. Kollintzas. "Stylized Facts of Business Cycles in the G7 from a Real Business Cycles Perspective." *European Economic Review* 38 (1994), 235–269.

[60] Foster, A.J. "Volume-Volatility Relationships for Crude Oil Futures Markets." *The Journal of Futures Markets* 15 (1995), 929-951.

[61] Frank, M. and T. Stengos. "Measuring the Strangeness of Gold and Silver Rates of Return." *Review of Economic Studies* 56 (1989), 553-567.

[62] French, K.R. "Detecting Spot Price Forecasts in Futures Prices." *Journal of Business* 59 (1986), S39-S54.

[63] Fuller, W.A. *Introduction to Statistical Time Series*. New York: John Wiley and Sons (1976).

[64] Gallant, A.R. and H. White. "On Learning the Derivatives of an Unknown Mapping with Multilayer Feedforward Networks." *Neural Networks* 5 (1992), 129-138.

[65] Gay, G.D. and Kim, T.H. "An Investigation into Seasonality in the Futures Market." *The Journal of Futures Markets* 7 (1987), 169-181.

[66] Garman, M.B. and M. Klass. "On the Estimation of Security Price Volatilities from Historical Data." *Journal of Business* 53 (1980), 67-78.

[67] Gencay, R. and W.D. Dechert. "An Algorithm for the n Lyapunov Exponents of an n-Dimensional Unknown Dynamical System." *Physica D* 59 (1992), 142-157.

[68] Ghashghaie, S., W. Breymann, J. Peinke, P. Talkner, and Y. Dodge. "Turbulent Cascades in Foreign Exchange Markets." *Nature* 381 (1996), 767-770.

[69] Grier, K.B., Ó.T. Henry, N. Olekalns, and K. Shields. "The Asymmetric Effects of Uncertainty on Inflation and Output Growth." *Journal of Applied Econometrics* 19 (2004), 551-565.

[70] Gonzalo, J. "Five Alternative Methods of Estimating Long Run Equilibrium Relationships." *Journal of Econometrics* 60 (1994), 203-233.

[71] Grammatikos, T. and A. Saunders. "Futures Price Variability: A Test of Maturity and Volume Effects." *Journal of Business* 59 (1986), 319-330.

[72] Haldane, A.G. and S.G. Hall. "Sterlings's Relationship with the Dollar and the Deutschemark: 1976-89." *The Economic Journal* 101 (1991), 436-443.

[73] Hamilton, J.D. "Oil and the Macroeconomy since World War II." *Journal of Political Economy* 91 (1983), 228–248.

[74] Hansen, P.R. and A. Lunde. "A Forecast Comparison of Volatility Models: Does Anything Beat a GARCH(1,1)?" *Journal of Applied Econometrics* 20 (2005), 829-889.

[75] Hiemstra, C. and J.D. Jones. "Testing for Linear and Nonlinear Granger Causality in the Stock Price-Volume Relation." *Journal of Finance* 49 (1994), 1639-1664.

[76] Hinich, M.J. "A Statistical Theory of Signal Coherence." *IEEE Journal of Oceanic Engineering* 25 (2000) 256-261.

[77] Hinich, M.J. "Testing for Caussianity and Linearity of a Stationary Time Series." *Journal of Time Series Analysis* 3 (1982), 169-176.

[78] Hinich, M.J. and A. Serletis. "Randomly Modulated Periodic Signals in Alberta's Electricity Market." *Studies in Nonlinear Dynamics and Econometrics* 10(3) (2006)), Article 5. (Reprinted in this volume as Chapter 19).

[79] Hinich, M.J. and P. Wild. "Testing Time-Series Stationarity Against an Alternative Whose Mean is Periodic." *Macroeconomic Dynamics* 5 (2001), 380-412.

[80] Hodrick, R.J. and E.C. Prescott. "Post-War U.S. Business Cycles: An Empirical Investigation." Working Paper, Carnegie Mellon University (1980).

[81] Hsieh, D.A. "The Statistical Properties of Daily Foreign Exchange Rates: 1974-1982." *Journal of International Economics* 24 (1988), 129-145.

[82] Ivanova, K. and M. Ausloos. "Low q-Moment Multifractal Analysis of Gold Price, Dow Jones Industrial Average and BGL-USD Exchange Rate." *The European Physical Journal B* 8 (1999), 665-669.

[83] Johansen, S. "Statistical Analysis of Cointegration Vectors." *Journal of Economic Dynamics and Control* (1988), 231-254.

[84] Johansen, S. and K. Juselius. "Maximum Likelihood Estimation and Inference on Cointegration with Applications to the Demand for Money." *Oxford Bulletin of Economics and Statistics* 52 (1990), 169-210.

[85] Johansen, S. and K. Juselius. "The Power Function for the Likelihood Ratio Test for Cointegration." In J. Gruber (ed.), *Econometric Decision Models: New Methods of Modelling and Applications*, Springer Verlag, New York (1991), 323-335.

[86] Johansen, S. and K. Juselius. "Some Structural Hypotheses in a Multivariate Cointegration Analysis of the Purchasing Power Parity and the Uncovered Interest Parity for the UK." *Journal of Econometrics* 53 (1992), 211-244.

[87] Kaplan, D.T. "Exceptional Events as Evidence for Determinism." *Physical D* 73 (1994), 38-48.

[88] Kearns, P. and A.R. Pagan. "Australian Stock Market Volatility: 1875-1987." *The Economic Record* 69 (1993), 163-178.

[89] Kenyon, D., K. Kling, J Jordan, W. Seala, and N. McCabe. "Factors Affecting Agricultural Futures Price Variance." *The Journal of Futures Markets* 7 (1987), 73-91.

[90] King, M. and M. Cuc. "Price Convergence in North American Natural Gas Spot Markets." *The Energy Journal* 17 (1996), 17-42.

[91] King, R.G. and S.T. Rebelo. "Low Frequency Filtering and Real Business Cycles." *Journal of Economic Dynamics and Control* 17 (1993), 207-231.

[92] Kroner, K.F. and V.K. Ng. "Modeling Asymmetric Comovements of Asset Returns." *The Review of Financial Studies* 11 (1998), 817-844.

[93] Kydland, F.E. and E.C. Prescott. "Business Cycles: Real Facts and a Monetary Myth." Federal Reserve Bank of Minneapolis *Quarterly Review* (Spring 1990), 3–18.

[94] Lamoureux, C. and W. Lastrapes. "Persistence in Variance, Structural Change, and the GARCH Model." *Journal of Business and Economic Statistics* 8 (1990), 225-234.

[95] Lee, J.H.H. "A Lagrange Multiplier Test for GARCH Models." *Economics Letters* 37 (1991), 265-271.

[96] León, A. and A. Rubia. "Testing for Weekly Seasonal Unit Roots in the Spanish Power Pool." In Derek W. Bunn (ed.), *Modelling Prices in Competitive Electricity Markets,* Wiley Series in Financial Economics (2004), 131-145.

[97] Li, T.H. and M.J. Hinich. "A Filter Bank Approach to Modeling and Forecasting Seasonal Patterns." *Technometrics* 44 (2002), 1-14.

[98] Li, W. "Absence of $1/f$ Spectra in Dow Jones Average." *International Journal of Bifurcation and Chaos* 1 (1991), 583-597.

[99] Liew, K.Y. and R.D. Brooks. "Returns and Volatility in the Kuala Lumpur Crude Palm Oil Futures Market." *The Journal of Futures Markets* 18 (1998), 985-999.

[100] Ljung, T. and G. Box. "On a Measure of Lack of Fit in Time Series Models." *Biometrica* 66 (1979), 66-72.

[101] Lucas, R.E., Jr. "Understanding Business Cycles." In K. Brunner and A.H. Meltzer (eds.), *Stabilization of the Domestic and International Economy*, Vol 5 of the Carnegie-Rochester Conference Series on Public Policy, North-Holland, Amsterdam, 1977, pp. 7-29.

[102] Ma, C.K., J.M. Mercer, and M.A. Walker. "Rolling Over Futures Contracts: A Note." *The Journal of Futures Markets* 12 (1992), 203-217.

[103] MacKinnon, J.G. "Critical Values for Cointegration Tests," Chapter 13 in R.F. Engle and C.W.J. Granger (eds.), *Long-Run Economic Relationships: Readings in Cointegration*, Oxford University Press 1991.

[104] MacKinnon, J.G. "Approximate Asymptotic Distribution Functions for Unit-Root and Cointegration Tests." *Journal of Business and Economic Statistics* 12 (1994), 167-176.

[105] Malick, W.M. and R.W. Ward. "Stock Effects and Seasonality in the FCOJ Futures Basis." *The Journal of Futures Markets* 7 (1987), 157-167.

[106] Mandelbrot B.B., *The Fractal Geometry of Nature*. Freeman, San Francisco (1982).

[107] Mantegna, R.N. and H.E. Stanley. "Turbulence and Financial Markets." *Nature* 376 (1996), 587-588.

[108] McCullough, R. "Spot Markets and Price Spreads in the Western Interconnection." *Public Utilities Fortnightly* 134 (1996), 32-37.

[109] Milonas, N.T. "Price Variability and the Maturity Effect in Futures Markets." *Journal of Futures Markets* 6 (1986), 444-459.

[110] Milonas, N.T. "Measuring Seasonalities in Commodity Markets and the Half-Month Effect." *The Journal of Futures Markets* 11 (1991), 331-345.

[111] Mork, K.A. "Oil and the Macroeconomy When Prices Go Up and Down: An Extension of Hamilton's Results." *Journal of Political Economy* 97 (1988), 740-744.

[112] Najand, M. and K. Yung. "A GARCH Examination of the Relationship between Volume and Price Variability in Futures Markets." *The Journal of Futures Markets* 11 (1991), 613-621.

[113] Nelson, D.B. "Conditional Heteroscedasticity in Asset Returns: A New Approach." *Econometrica* 59 (1991), 347-370.

[114] Nelson, C.R. and C.I. Plosser. "Trends and Random Walks in Macroecnomic Time Series: Some Evidence and Implications." *Journal of Monetary Economics* 10 (1982), 139-162.

[115] Ng, S. and P. Perron. "Estimation and Inference in Nearly Unbalanced Nearly Cointegrated Systems." *Journal of Econometrics* 79 (1997), 53-81.

[116] Nychka, D.W., S. Ellner, R.A. Gallant, and D. McCaffrey. "Finding Chaos in Noisy Systems." *Journal of the Royal Statistical Society B* 54 (1992), 399-426.

[117] Osterwald-Lenum, M. "Practitioners' Corner: A Note with Quantiles for the Asymptotic Distribution of the Maximum Likelihood Cointegration Rank Test Statistic." *Oxford Bulletin of Economics and Statistics* 54 (1992), 461-71.

[118] Pantula, S.G., G. Gonzalez-Farias, and W.A. Fuller. "A Comparison of Unit Root Test Criteria." *Journal of Business and Economic Statistics* 12 (1994), 449-59.

[119] Papaioannou, G. and A. Karytinos. "Nonlinear Time Series Analysis of the Stock Exchange: The Case of an Emerging Market." *International Journal of Bifurcation and Chaos* 5 (1995), 1557-1584.

[120] Parkinson, M. "The Extreme Value Method for Estimating the Variance of the Rate of Return." *Journal of Business* 53 (1980), 61-65.

[121] Perron, P. "Trends and Random Walks in Macroeconomic Time Series: Further Evidence From a New Approach." *Journal of Economic Dynamics and Control* 12 (1988), 297-332.

[122] Perron, P. "The Great Crash, the Oil Price Shock, and the Unit Root Hypothesis." *Econometrica* 57 (1989), 1361-1401.

[123] Perron, P. "Tests of Joint Hypotheses in Time Series Regression With a Unit Root." In G.F. Rhodes and T.B. Fomby, *Advances in Econometrics: Co-Integratation, Spurious Regression and Unit Roots*, Vol. 8, JAI Press (1990).

[124] Pesaran, M.H., Y. Shin, and R.J. Smith. "Bounds Testing Approaches to the Analysis of Long Run Relationships." *Journal of Applied Econometrics* 16 (2001), 289-326.

[125] Phillips, P.C.B. "Understanding Spurious Regression in Econometrics". *Journal of Econometrics* 33 (1986), 311-340.

[126] Phillips, P.C.B. "Time Series Regression with a Unit Root.' *Econometrica* 55 (1987), 277-301.

[127] Phillips, P.C.B. and P. Perron. "Testing for a Unit Root in Time Series Regression." *Biometrica* 75 (1988), 335-46.

[128] Pindyck, R.S. and J.J. Rotemberg. "The Excess Co-movement of Commodity Prices." *The Energy Journal* 100 (1990), 1173-1189.

[129] Plourde, A. and G.C. Watkins. "Relationships Between Upstream Prices of Canadian Crude Oil and Natural Gas." Discussion Paper, Faculty of Business, University of Alberta (August, 2000).

[130] Power Pool of Alberta Discussion Paper. "Assessment of Pool Price Deficiency Regulation and Intra Day Market – Next Steps." (October 4, 2001).

[131] Prescott, E.C. "Theorey Ahead of Business Cycle Measurement." Federal Reserve Bank of Minneapolis *Quarterly Revi*ew 10 (1986), 9-22.

[132] Provenzale, A., L.A. Smith, R. Vio, and G. Murane. "Distinguishing Between Low-Dimensional Dynamics and Randomness in Measure Time Series." *Physica D* 58 (1992), 431-491.

[133] RTO West. "RTO West Potential Benefits and Costs". Final Draft (October 23, 2000).

[134] Said, E. and D.A. Dickey. "Testing for Unit Roots in Autoregressive-Moving Average Models of Unknown Order." *Biometrica* 71 (1984), 599-607.

[135] Samuelson, P.A. "Proof that Properly Anticipated Prices Fluctuate Randomly." *Industrial Management Review* 6 (1965), 41-49.

[136] Scheinkman, J.A. and B. LeBaron. "Nonlinear Dynamics and Stock Returns." *Journal of Business* 62 (1989), 311-337.

[137] Schwartz, G. "Estimating the Dimension of a Model." *The Annals of Statistics* 6 (1978), 461-464.

[138] Serletis, A. "Rational Expectations, Risk and Efficiency in Energy Futures Markets." *Energy Economics* 13 (1991), 111-15 (Reprinted in this volume as Chapter 2).

[139] Serletis, A. "Maturity Effects in Energy Futures." *Energy Economics* 14 (1991), 150-57 (Reprinted in this volume as Chapter 3).

[140] Serletis, A. "Unit Root Behavior in Energy Futures Prices." *The Energy Journal* 13 (1992), 119-128 (Reprinted in this volume as Chapter 1).

[141] Serletis, A. "A Cointegration Analysis of Petroleum Futures Prices." *Energy Economics* 16 (1994), 93-97 (Reprinted in this volume as Chapter 5).

[142] Serletis, A. "Is There an East-West Split in North American Natural Gas Markets?" *The Energy Journal* 18 (1997), 47-62 (Reprinted in this volume as Chapter 6).

[143] Serletis, A. and I. Andreadis. "Nonlinear Time Series Analysis of Alberta's Deregulated Electricity Market." In Derek W. Bunn (ed.), *Modelling Prices in Competitive Electricity Markets,* Wiley Series in Financial Economics (2004), 147-159.

[144] Serletis, A. and I. Andreadis. "Random Fractal Structures in North American Energy Markets." *Energy Economics* 26 (2004), 389-399 (Reprinted in this volume as Chapter 18).

[145] Serletis, A. and D. Banack. "Market Efficiency and Cointegration: An Application to Petroleum Markets." *The Review of Futures Markets* 9 (1990), 372-80.

[146] Serletis, A. and P. Dormaar. "Imports, Exports, and Prices in Alberta's Deregulated Power Market." In W. David Walls (ed.) *Quantitative Analysis of the Alberta Power Market.* Van Horne Institute (2006).

[147] Serletis, A. and P. Gogas. "The North American Natural Gas Liquids Markets are Chaotic." *The Energy Journal* 20 (1999), 83-103 (Reprinted in this volume as Chapter 17).

[148] Serletis, A. and P. Gogas. "Chaos in East European Black-Market Exchange Rates." *Research in Economics* 51 (1997), 359-385.

[149] Serletis, A. and J. Herbert. "The Message in North American Energy Prices." *Energy Economics* 21 (1999), 471-483 (Reprinted in this volume as Chapter 13).

[150] Serletis, A. and V. Hulleman. "Business Cycles and the Behavior of Energy Prices." *The Energy Journal* 15 (1994), 125–134 (Reprinted in this volume as Chapter 4).

[151] Serletis, A. and T. Kemp. "The Cyclical Behavior of Monthly NYMEX Energy Prices." *Energy Economics* 20 (1998), 265-271 (Reprinted in this volume as Chapter 12).

[152] Serletis, A. and R. Rangel-Ruiz. "Testing for Common Features in North American Energy Markets." *Energy Economics* 26 (2004), 401-414 (Reprinted in this volume as Chapter 14).

[153] Serletis, A. and A. Shahmoradi. "Business Cycles and Natural Gas Prices." *OPEC Review* (2005), 75-84 (Reprinted in this volume as Chapter 7).

[154] Serletis, A. and A. Shahmoradi. "Futures Trading and the Storage of North American Natural Gas." *OPEC Review* (2006), 19-26 (Reprinted in this volume as Chapter 8).

[155] Shields, K., N. Olekalns, Ó.T. Henry, and C. Brooks. "Measuring the Response of Macroeconomic Uncertainty to Shocks." *Review of Economics and Statistics* 87 (2005), 362-370.

[156] Spiegel, M.R. *Statistics*. New York: McGraw-Hill (1988).

[157] Stock, J. "Asymptotic Properties of Least-Squares Estimators of Cointegrating Vectors." *Econometrica* 55 (1987), 1035-56.

[158] Stock, J. and M.W. Watson. "Testing for Common Trends". *Journal of the American Statistical Association* 83 (1988), 1097-1107.

[159] Summers, L.H. "Does the Stock Market Rationally Reflect Fundamental Values." *Journal of Finance* 41 (1986), 591-601.

[160] Telser, L.G. "Futures Trading and the Storage of Cotton and Wheat." *Journal of Political Economy* 66 (1958), 233-55.

[161] Turvey, R. "The Economics of Interconnectors." Unpublished paper.

[162] Turvey, R. "The Challenges of Securing the Best Possible Use of Transmission Interconnections." *CERI 2005 Electricity Conference* (2005).

[163] Vahid, F. and R.F. Engle. "Common Trends and Common Cycles." *Journal of Applied Econometrics* 8 (1993), 341-360.

[164] Vassilicos, J.C. and J.C.R. Hunt. "Fractal Dimensions and Spectra of Interfaces with Application to Turbulence." *Proceedings of the Royal Society of London Series A* 435 (1991), 505-534.

[165] Vassilicos, J.C., A. Demos, and F. Tata. "No Evidence of Chaos But Some Evidence of Multifractals in the Foreign Exchange and the Stock Markets." In *Applications of Fractals and Chaos*, A.J. Crilly, R.A. Earnshaw, and H. Jones (eds.) Springer-Verlag, New York (1993), pp. 249-265.

[166] White, H. "Some Asymptotic Results for Learning in Single Hidden-Layer Foodforward Network Models." *Journal of the American Statistical Association* 84 (1989), 1003-1013.

[167] Woo, C.K., D. Lloyd-Zanetti, and I. Horowitz. "Electricity Market Integration in the Pacific Northwest." *The Energy Journal* 18 (1997), 75-101.

[168] Working, H. "The Theory of the Price of Storage." *American Economic Review* 39 (1949), 1254-1262.

[169] Zivot, E. and D.W.K. Andrews. "Further Evidence on the Great Crash, the Oil Price Shock, and the Unit Root Hypothesis." *Journal of Business and Economic Statistics* 10 (1992), 251-270.

Author Index

Topic Index

Printed in the United States
By Bookmasters